T0336090

# Modeling and State Estimation of Automotive Lithium-Ion Batteries

This book aims to evaluate and improve the state of charge (SOC) and state of health (SOH) of automotive lithium-ion batteries.

The authors first introduce the basic working principle and dynamic test characteristics of lithium-ion batteries. They present the dynamic transfer model, compare it with the traditional second-order reserve capacity (RC) model, and demonstrate the advantages of the proposed new model. In addition, they propose the chaotic firefly optimization algorithm and demonstrate its effectiveness in improving the accuracy of SOC and SOH estimation through theoretical and experimental analysis.

The book will benefit researchers and engineers in the new energy industry and provide students of science and engineering with some innovative aspects of battery modeling.

**Shunli Wang** is Professor, Academic Dean, Academic Leader of the National Electrical Safety and Quality Testing Center, Academician of the Russian Academy of Natural Sciences, Senior Overseas Study Talent of Sichuan Province, and Academic and Technical Leader of China Science and Technology City. He is an authoritative expert in renewable energy research and is interested in modeling, state estimation, and safety management for energy storage systems.

**Fan Wu** is in the School of Information Engineering at Southwest University of Science and Technology. Her research direction is the high-precision state estimation and safety management for energy storage systems. She has published two SCI papers and one international conference paper, authorized a software copyright, and won a national scholarship.

**Jialu Qiao** is in the School of Information Engineering at Southwest University of Science and Technology. Her research direction is the estimation of the state of charge and health of lithium-ion batteries. She has published three SCI papers as the lead author and one international conference paper, authorized software copyright, and applied for a national invention patent.

**Carlos Fernandez** is Professor at Robert Gordon University, Head of the MSc in Analytical Sciences Program, and Member of the National Institute of Battery Research. Specializing in analytical chemistry/sensors and materials/renewable energy research, he has led and participated in more than 20 projects with a cumulative economic benefit of £270,000. He has published over 150 papers in *SCI-1* and other journals, participated in over 200 reviews, and published three books.

**Josep M. Guerrero** has been Professor in the Department of Energy Technology at Aalborg University in Denmark since 2011, where he leads the Microgrid research project. In 2019, he became Fluff Fellow at the Villum Fonden, where he is the Founder and Director of the Center for Research on Microgrids (CROM). His research interests are oriented toward different microgrid aspects, including power electronics distributed energy storage systems, hierarchical and cooperative control, energy management systems, smart metering, sustainable energy, and the internet of things for AC/DC microgrid clusters and islanded mini-grids.

# Modeling and State Estimation of Automotive Lithium-Ion Batteries

Shunli Wang, Fan Wu, Jialu Qiao,
Carlos Fernandez, and Josep M. Guerrero

CRC Press is an imprint of the
Taylor & Francis Group, an **informa** business

MATLAB® and Simulink® are trademarks of The MathWorks, Inc. and are used with permission. The MathWorks does not warrant the accuracy of the text or exercises in this book. This book's use or discussion of MATLAB® or Simulink® software or related products does not constitute endorsement or sponsorship by The MathWorks of a particular pedagogical approach or particular use of the MATLAB® and Simulink® software.

First edition published 2025
by CRC Press
2385 NW Executive Center Drive, Suite 320, Boca Raton FL 33431

and by CRC Press
4 Park Square, Milton Park, Abingdon, Oxon, OX14 4RN

*CRC Press is an imprint of Taylor & Francis Group, LLC*
© 2025 Shunli Wang, Fan Wu, Jialu Qiao, Carlos Fernandez and Josep M. Guerrero

Reasonable efforts have been made to publish reliable data and information, but the author and publisher cannot assume responsibility for the validity of all materials or the consequences of their use. The authors and publishers have attempted to trace the copyright holders of all material reproduced in this publication and apologize to copyright holders if permission to publish in this form has not been obtained. If any copyright material has not been acknowledged please write and let us know so we may rectify in any future reprint.

Except as permitted under U.S. Copyright Law, no part of this book may be reprinted, reproduced, transmitted, or utilized in any form by any electronic, mechanical, or other means, now known or hereafter invented, including photocopying, microfilming, and recording, or in any information storage or retrieval system, without written permission from the publishers.

For permission to photocopy or use material electronically from this work, access www.copyright.com or contact the Copyright Clearance Center, Inc. (CCC), 222 Rosewood Drive, Danvers, MA 01923, 978–750–8400. For works that are not available on CCC please contact mpkbookspermissions@tandf.co.uk

*Trademark notice*: Product or corporate names may be trademarks or registered trademarks and are used only for identification and explanation without intent to infringe.

ISBN: 978-1-032-77791-7 (hbk)
ISBN: 978-1-032-78115-0 (pbk)
ISBN: 978-1-003-48625-1 (ebk)

DOI: 10.1201/9781003486251

Typeset in Minion
by Apex CoVantage, LLC

# Contents

# Preface

AT PRESENT, THE ECONOMY is in a stage of rapid growth, and the demand for fossil energy is increasing. Therefore, the threat of energy depletion is earlier and more severe than before. In recent years, new energy has developed very fast in the automotive field. Under the dual problems of economic energy shortage and the deteriorating environment, many countries have spent a lot of financial and human resources on the research and popularization of new energy. The withdrawal of traditional fuel vehicles has become an inevitable trend, and the development of electric cars will be unstoppable.

The onboard power lithium-ion battery is the core technical problem of further technological breakthroughs and broad application of new energy vehicles. To realize the real-time accurate estimation of state of charge (SOC) and state of health (SOH) is of great significance to strengthen the real-time monitoring function of the power battery management system and ensure the safe and reliable operation of power batteries.

In this book, taking the ternary lithium-ion battery as the research object, the battery charge and discharge experiments under complex working conditions are analyzed, and a dynamic battery migration model that can accurately describe the characteristics of the battery is built. The double filter is constructed, and the population intelligent optimization of the firefly algorithm is introduced to achieve a powerful high-precision SOC and SOH collaborative estimation of lithium-ion batteries. These improved methods effectively provide real-time and accurate residual power status and health status, extend the service life of the power lithium-ion battery, and provide a solid guarantee for the safety of electric vehicles.

The main research contents and chapters of this subject are divided as follows: The first chapter introduces the background and significance of this research topic in detail and introduces the current status of SOC estimation, SOH estimation, and SOC/SOH co-estimation. The second chapter is the experimental characteristics analysis of the power lithium-ion battery. The third chapter is the dynamic migration model construction of a power lithium-ion battery. The fourth chapter is a co-estimation study of SOC and SOH for power lithium-ion batteries based on the firefly optimization algorithm. This chapter discusses the progressive reciprocal correction of SOC and SOH. Chapter 5 is the experimental verification and analysis of the co-estimation of SOC and SOH. Chapter 6 is the summary and prospect of the work on this topic. The work and the results of the institute are summarized, the shortcomings in the current research methods are analyzed, and the subsequent research work of this book is discussed.

CHAPTER 1

# Introduction

THE POWER LITHIUM-ION BATTERY has advantages like light weight, high energy density, and low pollution and has become the primary source of new energy vehicles. Lithium-ion battery state of charge and charge of health as new energy vehicles in the work process must accurately monitor the two core states. Their accurate estimation for the timely provision of exact power state and car safety is of great significance.

## 1.1 RESEARCH BACKGROUND AND SIGNIFICANCE

The birth of the automobile is an essential symbol of the progress and development of modern industrial civilization, which provides great convenience for the production of society and human life. With global energy and environmental and ecological problems becoming increasingly prominent, the shortage of coal and other resources and the urgency of air pollution all promote the further application of clean energy in the automotive industry (Guanghui Zhao, Wang, and Chen 2022; Q. Zhao, Hu, et al. 2020; Zou et al. 2020). The Chinese government is committed to promoting the healthy development of the new energy vehicle industry. It has promulgated a series of relevant policies and regulations to improve the industrialization system of new energy vehicles gradually.

In the new energy vehicles (J. Mao et al. 2020; Q. Tan et al. 2023; Hao Zhang and Cai 2020) widely used at present, power batteries are often used as the primary power source of the car. Compared with traditional fuel vehicles, the pollution caused to the environment is much less, and it has a wide range of development prospects. Lithium-ion batteries (L. Zhang, Li, et al. 2021; Kheirkhah-Rad et al. 2023; G. Huang et al. 2021;

DOI: 10.1201/9781003486251-1

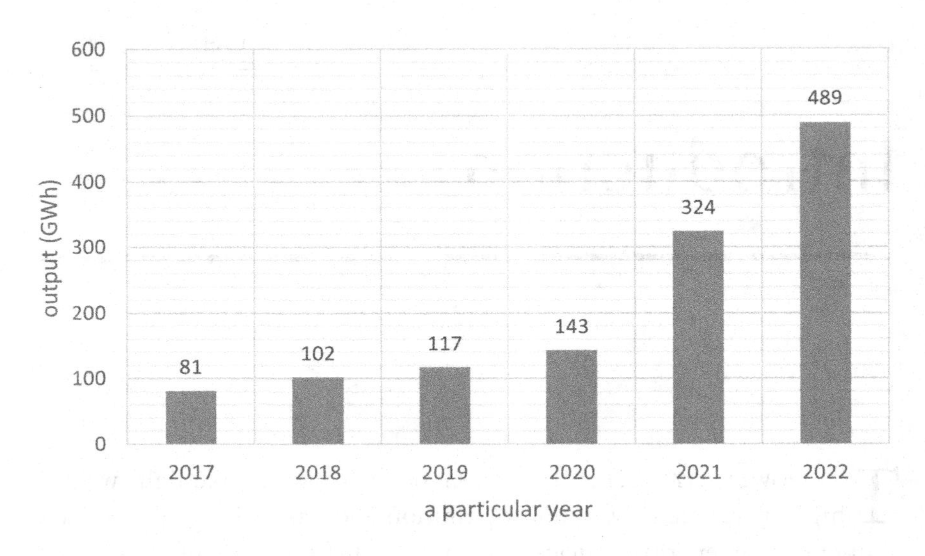


FIGURE 1.1 Power lithium-ion battery production in China.

Degen et al. 2023), because of their light weight, low self-discharge rate, high energy density, and low pollution advantage, have become the essential power source of pure electric vehicles (H. Chen et al. 2020; Q. Chen et al. 2022; Fraile Ardanuy et al. 2022; Gutsch and Leker 2022; Leon and Miller 2020). The production of power lithium-ion batteries in China from 2017 to 2022 is shown in Figure 1.1.

As can be seen from Figure 1.1, China's lithium-ion battery production will reach 324 GWh in 2021, with a year-on-year growth of 126.57%. In 2022, lithium-ion battery production in China will get 489 GWh, making a qualitative leap. Lithium-ion battery industry innovation in China continues to accelerate, the supply capacity of high-end products has been continuously improved, and energy density has been made in ternary batteries and lithium-ion iron phosphate batteries (G. Huang et al. 2021; C. Sun et al. 2020; Mohan et al. 2022; Na et al. 2021; Schlesinger et al. 2021; Tabelin et al. 2021). Research and development in China on new energy vehicle lithium-ion battery technology is constantly strengthened. Major manufacturers actively research and develop new materials, processes, and systems to improve the performance of all aspects of the battery. The NCM 811 battery (M.Y. Kim et al. 2022) of the Ningde era has achieved large-scale application, with an energy density of more than 250 Wh/kg, which is more than 30% higher than the traditional NCM523 battery (Shin et al. 2020). China's new energy vehicle lithium-ion battery industry chain is also constantly improving and has formed a complete industrial chain, from raw material procurement, battery production, and system integration to recycling and so on (Y. Miao, Liu, et al. 2023; Zeng et al. 2023; Yue et al. 2021; D. Zhang, Tan, et al. 2022; Zhi et al. 2022). Among them, Ningde Times, BYD, CATL, and other enterprises have become the world's leading battery manufacturers, and their battery products have been widely used in the world.

As a significant judge of technical difficulty in the development process of new energy vehicles, the performance of power lithium-ion batteries often directly affects whether the user's experience of driving new energy vehicles is good (G. Huang et al. 2021; Kacica 2020; Kheirkhah-Rad et al. 2023; Paul et al. 2020). If the battery is wrong, it will cause a large loss, and the cost of maintenance and replacement is much higher than the fuel consumption cost of the fuel car (Z. Duan et al. 2021; J. Han et al. 2022; Huiqian Sun, Jing, et al. 2023; Haoyi Zhang, Zhao, et al. 2021). Secondly, lithium-ion batteries are affected by environmental factors, among which temperature and aging are two key factors, resulting in the performance of new energy vehicles in the cold winter being far less than that of fuel vehicles (Aylagas et al. 2022; D. Li, Li, et al. 2022; Jiejia Wang, Jia, and Zhang 2022; Shujie Wu et al. 2020; Jingjing Xu, Cai, et al. 2023). If the battery is used for a long time, the case aging is also prone to failure. In addition, if the owner of the car excessively charges and discharges the battery (Tsai and Peng 2023; Yun Yang et al. 2020; Yang Yang, Lan, et al. 2022; Y. Ye, Zhang, et al. 2023; Q. Yuan et al. 2020; Guanghui Zhao, Wang, and Chen 2022), or when long driving causes battery overload, it is likely to cause the adverse consequences of thermal runaway, which will directly affect the life and property safety of the driver.

To ensure the safe and reliable operation of onboard lithium-ion batteries in new energy vehicles under complex driving conditions, an efficient and accurate battery management system (BMS) is indispensable (S. Li, He, and Li 2019; Yong Chen, Li, et al. 2021; Sung et al. 2016; J. Yang, Liu, and Duan 2020). The BMS can provide real-time feedback and prediction of the battery state and evaluate the current performance status of the battery (Bustos et al. 2023; Jiao et al. 2023; Xiaoyu Li, Xu, et al. 2023; Yuefeng Liu, He, et al. 2022; Navega Vieira et al. 2022). The accurate prediction of the current battery power and health characteristics can ensure that the battery works in the defined safe work zone, can realize the comprehensive monitoring and management of the vehicle lithium-ion battery, can help users optimize battery use, prolong battery life, and reduce the use cost (L. Wu et al. 2021; Mohammed and Saif 2021; Yuhong Chen 2022; Ziegler et al. 2021). The SOC (Cui, Kang, et al. 2022a; Yuefeng Liu, Li, et al. 2021; B. Du et al. 2021; F. Wu, Wang, et al. 2023) and SOH (S. Yang et al. 2021; L. Hu, Wang, and Ding 2023; Yuanyuan Jiang, Zhang, et al. 2020; H. Ji et al. 2020) are the two most critical states of lithium-ion batteries, which BMS must accurately assess. SOC is used to represent the remaining available capacity of lithium-ion batteries, generally referring to the remaining power of a battery. It is also a representation of the charging state of the battery, and its accurate evaluation can help the user understand the charging state of the battery, judge whether the battery meets the needs, and decide whether it needs to be charged (X. Wei, Mo, and Feng 2019). In addition, the accurate estimation of SOC can also optimize the battery use and charging strategy for users, improve the battery use efficiency, and extend the battery life.

After repeated charging and discharge of the battery, it will face the risk of aging, that is, the performance of the battery will gradually decrease with the increase of its use time and the increase of its aging degree (S. Wang, Wu, et al. 2023; Bao and Gong 2023; Lei Feng et al. 2021; Fluegel et al. 2022). Generally speaking, when the SOH of the battery is below 80% (C. Jiang, Wang, et al. 2020; Iurilli et al. 2022; Luo et al. 2023; D. Ouyang, Weng, et al. 2022, 2023), the failure threshold is reached, and the battery can no longer be used normally. SOH is the state quantity used to reflect the degree of aging and decay of the battery. The accurate evaluation of SOH can help drivers know the health of the battery of the car they drive in real-time, to remind them to replace the battery in time, to avoid accidents that endanger the driving safety caused by the battery not maintaining the regular operation under the performance (Q. Zhang, Huang, et al. 2022; Bao and Gong 2023; S.-J. Park et al. 2023; C.-j. Wang, Zhu, et al. 2020). In addition, its accurate estimation can also optimize battery use and charging strategies, extend battery life, and reduce use costs.

In conclusion, the estimation of SOC and SOH is crucial for battery management and use and can help users understand the status and health of batteries. It can optimize the battery use and charging strategies, prolong the battery life, improve the energy efficiency of batteries, and reduce the use cost. The accurate evaluation of SOC/SOH of power lithium-ion batteries is of great significance to the improvement of the overall performance of the whole life cycle of batteries and the driving safety of new energy vehicles.

## 1.2 RESEARCH STATUS AT HOME AND ABROAD

### 1.2.1 Current Status of SOC Estimation

At present, the SOC generally recognized by relevant researchers at home and abroad is defined as the ratio of the battery's remaining power under the standard discharge rate condition and the rated capacity under the same situation (Yanbo Che et al. 2021; Cui, Wang, et al. 2022; X. Duan 2023; Huo et al. 2023), which is defined as shown in Equation (1.1).

$$SOC = \frac{Q_r}{Q_n} \times 100\% \tag{1.1}$$

In Equation (1.1), $Q_r$ is the remaining battery power and $Q_N$ is the rated capacity under the same conditions. SOC = 100% means that the battery is fully charged at the moment. If the battery is still charged at this time, it may explode due to overcharging (Kawahara et al. 2021; Huan Li, Zou, et al. 2021; Jiabo Li, Ye, et al. 2023; Renzheng Li, Hong, et al. 2022). SOC = 0% indicates that the battery is fully discharged at the moment. If the battery is still discharged at this time, it may be damaged due to over-discharge. According to the existing studies, the methods of SOC estimation can be divided into three categories: traditional estimation, data-driven, and model-based methods (Q. Lin et al. 2022; Peng et al. 2022; W. Ren, Zheng, et al. 2022). The traditional estimation method includes the ampere time integral method and the open circuit voltage method (S. Wang, Takyi-Aninakwa, Jin, et al. 2022; Jiaqiang et al. 2022; Lei Feng et al. 2021). An integral method calculation is very simple and easy to understand. The current value in the process of charge and discharge integral can accumulate battery charge and discharge power and, according to the SOC calculation formula, can directly find the current value. However, the method is heavily dependent on the initial SOC value, and the accumulation of discharge charge will produce an error accumulation phenomenon (Sonwane 2020; S. Wang, Takyi-Aninakwa, Yu, et al. 2022; Y. Xie, Li, Hu, et al. 2023). Jiang et al. (Jiaqiang et al. 2022) proposed an extended Kalman filter (EKF) algorithm to estimate the SOC based on the ampere-hour integral and open circuit voltage. Then, the Simscape battery model is established to estimate battery parameters, and the HPPC experiment is employed to verify the SOC estimation precision. Moreover, based on the battery SOC inconsistency laws, the battery equalization control strategy is presented. Finally, the experimental bench is set up to validate the effectiveness of the equalization strategies. The experimental results show that the maximum SOC estimation error is about 4%, so this SOC estimation method meets the accuracy requirement. These results indicate that active equalization control in this chapter cannot only improve cell inconsistency but also improve the energy utilization of the battery pack in the process of charging and discharging. The open circuit voltage method is based on the relationship between the open circuit voltage of the lithium-ion battery at different SOC. It estimates the SOC of the battery by measuring the open circuit voltage of the battery (Shan et al. 2022; L.-L. Li, Liu, and Wang 2020). Since the open circuit voltage of the lithium-ion battery changes with the SOC, the SOC of the battery can be inferred by measuring the open circuit voltage of the battery. Specifically, the battery can be discharged first to reduce the SOC to a certain extent. Then, the battery can be left in the open state for a while, waiting for the chemical reaction of the battery to reach balance (R. Xiao et al. 2023; Y. Xie et al. 2022; F. Yang, Shi, et al. 2023; Yun and Kong 2022). At this time, the open circuit voltage of the battery can be stabilized. According to the known relationship between the open circuit voltage and the SOC, the SOC of the battery can be introduced (Barai et al. 2015). According to Yuan et al. (B. Yuan et al. 2022), to obtain accurate SOC, the relationship between OCV and SOC requires to be in real time and accurate. Due to the difference in lithium-ion concentration and battery internal resistance in the lithium-ion battery, OCV has the characteristics of relaxation. It is necessary to study the relaxation behavior of battery OCV. They studied the OCV behavior and focused on the relationship between the time constant and polarization resistance with SOC during relaxation. The results show that when the SOC is 30%–100%, the time constant and polarization resistance of lithium-ion batteries are the smallest, the performance is the most stable, and the SOC estimation accuracy is the highest. This method takes a lot of time and is not suitable for online SOC estimation, let alone practical engineering applications.

More and more scholars are studying data-driven methods to estimate SOC, which builds neural network models based on the collected parameter data, such as current, voltage, and temperature (Z. Du, Zuo, et al. 2022; O. Ali, Ishak, et al. 2022; W. Bai et al. 2022; Junxiong Chen, Feng, et al. 2021; Jianlong Chen, Zhang, et al. 2023; F. Wu et al.). The input layer accepts the real-time data of the battery, the output layer outputs the SOC estimate of the battery, and the hidden layer carries out data processing and feature extraction (M. Hu et al. 2018; Cui, Kang, et al. 2022b; Z. Du, Zuo, et al. 2022; Fahmy et al. 2021). The neural network model (Xiaobo Zhao, Jung, et al. 2023; N. Ouyang, Zhang, et al. 2023; Premkumar et al. 2022; Ragone et al. 2021; Rimsha et al. 2023) was trained to estimate the SOC of the cell accurately. In training, it is necessary to take the known SOC value of the battery as the training target. The real-time data of the battery should be input into the model for training (Falai et al. 2022; Q. Gong, Wang, and Cheng 2022; S. Guo and Ma 2023; J. Hu et al. 2022; Jerouschek et al. 2020). By constantly trying and adjusting the optimization of the model of the parameters involved in the model, the output of the model can be as close to the actual value as possible. It also presents a solid versatility for many types of batteries, and universality, fault tolerance, and real-time are better (Lian et al. 2023; S.W. Park, Lee, and Won 2022; Y. Xie, Wang, et al. 2023). However, the scope of the application is limited by training data and high time cost, which is difficult to achieve in BMS. The radial base neural network (RBFNN) (Ghanbari-Adivi et al. 2022; Saeed Alzaeemi and Sathasivam 2021) was selected for model construction. Character extraction and mapping of input data are performed through a set of radial basis functions, and the model is accurate by constantly adjusting the parameters and weights of radial basis functions. In Cui, Kang, et al. (2022b), to obtain accurate SOC values, a hybrid method to achieve stable and real-time battery SOC estimation at different temperatures is composed of an improved bidirectional gated recurrent unit (BiGRU) network and unscented Kalman filtering (UKF). The proposed method is experimentally validated using data from UDDS and US06 driving cycles. The verification results show that the method can adapt to various working conditions and obtain good estimation accuracy and robustness, with MAE and RMSE less than 0.83% and 1.12%, respectively. After transfer learning, the method can also be applied to new lithium-ion batteries and achieve good estimation performance at new temperature conditions. The maximum errors are 4.98% and 5.76% at 25°C and −10°C, respectively. Therefore, the BiGRU-UKF method can achieve a more accurate and stable SOC estimation with good expansion performance for different lithium-ion batteries. Recently, with the continuous development of graphic processing unit (GPU) computing power, there has been an increasing interest in applying deep learning as SOC estimation approaches. Li et al. (Jiarui Li, Huang, Tang, et al. 2023) proposed a novel SOC estimation scheme for a lithium-ion energy storage system based on a convolutional neural network and long short-term memory

(CNN-LSTM) neural network. This method is completely driven by the actual operating data from a photovoltaic energy storage system without using any artificial battery models or inference systems. Compared with traditional SOC estimation methods, the CNN-LSTM model can overcome the deviation in estimation caused by voltage jump at the end of charge and discharge and provide satisfied SOC estimation results during the stabilized stage and various charging/discharging stages of the assembled lithium-ion batteries in the system. The calculation results indicate that this method enables fast and accurate SOC estimation, with an RMSE of less than 0.31% over the entire operating data of the photovoltaic energy storage system for a full day. Although the model has a good ability to solve nonlinear problems, it fails to better adapt to the complex and changeable actual driving conditions (Jiarui Li, Huang, Tang, et al. 2023; Lipu et al. 2020; Xingtao Liu, Li, et al. 2023; Luciani et al. 2021; L. Mao, Hu, Zhao, et al. 2023). The estimation accuracy cannot meet the actual requirements.

The model-based estimation method is the most common and widely used algorithm at present, which is usually combined with various filtering algorithms to realize the estimation of state quantity (Jo, Jung, and Roh 2021; S.-J. Kim et al. 2021; Bohao Li and Hu 2022; Huan Li, Wang, et al. 2022; Huan Li, Zou, et al. 2021). The filtering algorithm is the basis for the application of the battery model. The Kalman filter (KF) (Lange 2022; H. Liu, Hu, Su, et al. 2020; Xing Zhang, Yan, and Chen 2022) algorithm is an estimation method based on the state space model, which is one of the most commonly used algorithms for SOC estimation (F. Liu, Yu, et al. 2023; O. Rezaei, Habibifar, and Wang 2022; H. Wang, Zheng, and Yu 2021; Yuefei Wang, Huang, et al. 2021; Jieyu Xu and Wang 2022). It forms a matrix of the estimated states, such as SOC, and polarization voltage as a state matrix. The prior estimate of SOC is obtained according to the state space equation. Then, the preceding estimate is corrected and combined with the observed value, such as open circuit voltage, to get the assessment of the next moment (Yueshuai Fu and Fu 2023; J. Hong, Ramirez-Mendoza, and Lozoya-Santos 2020; Wenqian Li, Yang, et al. 2020; B. Liu 2020). Literatures (Al-Gabalawy et al. 2021; Gu et al. 2022; Qian et al. 2022; Liang Feng, Ding, and Han 2020) established the Thevenin equivalent circuit model, based on the recursive principle of extended Kalman filter (EKF) algorithm, an integration method used for SOC estimation correction, realize the effective combination of the two, in the case of currency fluctuation is still can recognize the online estimation of SOC, the estimation error can be controlled within 3%. According to Shi et al.

(N. Shi et al. 2022), for the problems of selection of forgetting factor and poor robustness and susceptibility to the noise of extended Kalman filtering algorithm, this chapter proposes an SOC estimation method for the lithium-ion battery based on adaptive extended Kalman filter using improved parameter identification. Firstly, the Thevenin equivalent circuit model is established, and the recursive least squares with forgetting factor (FFRLS) method is used to achieve the parameter identification. Secondly, an evaluation factor is defined, and fuzzy control is used to realize the mapping between the evaluation factor and the correction value of the forgetting element to realize the adaptive adjustment of the forgetting factor. Finally, the noise adaptive algorithm is introduced into the extended Kalman filtering algorithm (AEKF) to estimate the SOC based on the identification results, which is applied to the parameter identification the next time and executed circularly to realize the accurate estimation of SOC. The experimental results show that the proposed method has good robustness and analysis accuracy compared with other filtering algorithms under different working conditions, SOH, and temperature. Shrivastava et al. (Shrivastava et al. 2021) suggested the new dual forgetting factor-based adaptive extended Kalman filter (DFFAEKF) for SOC estimation. For all the considered test battery cells, the experimental results indicated that the combined SOC and SOE estimation method using the proposed DFFAEKF can estimate the battery states under dynamic operating conditions with root mean square error (RMSE) less than 0.85% and 0.95%, respectively. The proposed method also demonstrates fast convergence to its actual value under erroneous initial conditions.

### 1.2.2 Current Status of SOH Estimation

The aging of lithium-ion batteries will lead to the decline of their capacity, internal resistance, and other electrical performance, which affect the health of the battery (Z. Lin et al. 2023; F. Liu, Liu, et al. 2021; Y. Jia et al. 2023; N. Jiang and Pang 2022; Lai et al. 2022). Capacity attenuation is one of the main manifestations of lithium-ion battery aging, and its degree is directly proportional to SOH. When the SOH of the battery drops to a certain extent, the available capacity of the battery will not be able to meet the actual demand (S. Shen et al. 2021; Bing Li 2020; C. Liu, Wen, et al. 2022; L. Mao, Yang, et al. 2023; Hanlei Sun, Yang, et al. 2022). The SOH of the battery can be evaluated by measuring the existing capacity and the original capacity, by calculating the capacity decay rate of the storm, or by calculating the original and the real internal resistance of the storm

(Huiqin Sun, Wen, et al. 2022; Teng et al. 2023; Z. Xia and Abu Qahouq 2021; Jun Xu, Mei, et al. 2021). When SOH is 100%, it means that the battery is still in a new state. When SOH is reduced to 80%, the power battery is deemed to have been scrapped.

When battery capacity attenuation is taken as the evaluation index, SOH is often defined as the ratio of battery residual capacity and rated capacity (L. Yao et al. 2022; Yin et al. 2022; Xu Zhao, Hu, et al. 2023), which is defined as shown in Equation (1.2).

$$SOH = \frac{Q_{now}}{Q_{new}} \times 100\% \tag{1.2}$$

In the preceding equation, $Q_{new}$ represents the rated capacity of the battery in a new state, and $Q_{now}$ is the capacity of the current moment at which the battery is used after cyclic charge and discharge and is usually measured by the capacity calibration experiment. When the battery's internal resistance is increased as the evaluation index, SOH is defined as shown in Equation (1.3).

$$SOH = \frac{R_{aged} - R_{now}}{R_{aged} - R_{new}} \times 100\% \tag{1.3}$$

In the preceding equation, $R_{new}$ represents that the battery is not in use, $R_{aged}$ for the ohmic internal resistance when the battery health state is below 80% and reaches the failure threshold, and $R_{now}$ is ohm internal resistance at the current moment after the battery has been charged and discharged.

The research methods of SOH estimation at home and abroad are mainly divided into three directions: experimental test method, data-driven method, and model-based method. The experimental test method includes the complete discharge method and partial discharge method (G. Lee, Kim, and Lee 2022; H. Yao et al. 2020; X. Sun, Zhang, et al. 2023; Yalong Yang, Chen, et al. 2023). The principle is to measure the discharge time and discharge capacity of the battery under the condition of constant current discharge to calculate the SOH of the battery. This method is suitable for various types of batteries. In Yumeng Fu et al. (2022), to solve the problems of relevant research articles, which are primarily based on battery external information, such as current, voltage, and temperature, which are susceptible to fluctuation and ultimately affect the SOH estimation accuracy, the authors proposed a fast impedance calculation-based battery SOH estimation method for lithium-ion battery from the perspective of electrochemical impedance spectroscopy (EIS). In B. Jiang et al. (2022), the study investigated a systematic comparative study of three categories of features extracted from battery electrochemical impedance spectroscopy (EIS) in SOH estimation. The three representative features are broadband EIS feature, model parameter feature, and fixed-frequency impedance feature. Based on the deduced EIS features, a machine learning technique using Gaussian process regression is adopted to estimate battery SOH. Although the complete discharge method is simple and easy, it also has some limitations and disadvantages. For example, this method takes a long time and considerable electrical energy and is not suitable for online monitoring of battery health status (Qu et al. 2019; Braco et al. 2022; Hammou et al. 2023; X. Hu et al. 2020). Meanwhile, in practice, the setting of battery discharge time and discharge cut-off voltage can also affect the accuracy of the measurement results. The principle of the partial discharge method is to discharge the battery to a certain extent and then estimate its capacity according to the change in discharge time and battery voltage (C. Zhang, Zhao, et al. 2022; X. Zhu et al. 2022). Compared with the complete discharge method, the partial discharge method saves more time and energy and is suitable for online monitoring of battery capacity. This method needs reasonable setting and analysis according to the battery type and specification, facing great technical difficulties and poor applicability.

The advantages and disadvantages of the estimation of SOH are similar to the estimation of SOC (Bian et al. 2022; C. Chang et al. 2021; Yanbo Che, Cai, et al. 2022; C. Du et al. 2023). Its basic idea is, according to recorded data in the working state of the battery health state model, and then using the model to estimate the battery SOH, the model can use machine learning, artificial neural network establishment, and optimization to improve the estimation accuracy and stability (Cen and Kubiak 2020; Gao et al. 2022; P. Hu, Tang, et al. 2023). The neural network method is a commonly used data-driven method, and its basic idea is to use historical data to train network parameters (J. Tian et al. 2022; Ungurean, Micea, and Carstoiu 2020; Van and Quang 2023). When training the network, a large amount of battery working state data and the corresponding SOH data are needed to improve the accuracy of the network (Y. Gong et al. 2022; P. Hu, Tang, et al. 2023; Q. Lin, Xu, and Lin 2021; Ma et al. 2022; Meng, Geng, and Han 2023; H. Tian et al. 2020). When using the network for estimation, it is necessary to input the current battery operating state data and then output the corresponding

SOH value. The process can be estimated and updated using historical data and real-time data (Kawahara et al. 2021; J. Kim and Kowal 2021, 2022; Kurzweil, Frenzel, and Scheuerpflug 2022). The estimation of battery SOH by this method mainly includes two processes, namely, the learning process and the working process. The state-of-health (SOH) estimation is a challenging task for lithium-ion batteries, which contributes significantly to maximizing the performance of battery-powered systems and guiding battery replacement. Fan et al. (Yaxiang Fan et al. 2020) argue that the complexity of degradation mechanisms makes it possible for data-driven approaches to replace mechanism modeling approaches to estimate SOH. The proposed approach is based on a hybrid neural network called gate recurrent unit-convolutional neural network (GRU-CNN), which can learn the shared information and time dependencies of the charging curve with deep learning technology. Then, the SOH could be estimated with the newly observed charging curves, such as voltage, current, and temperature. Yang et al. (N. Yang, Song, et al. 2022) utilize the convolutional neural network (CNN) to extract indicators for both SOH and changes in SOH (Delta SOH) between two successive charge/discharge cycles. The random forest algorithm is then adopted to produce the final SOH estimate by exploiting the indicators from the CNNs. Performance evaluation is conducted using the partial discharge data with different SOC ranges created from a fast-discharging dataset. The proposed approach is compared with (1) a differential analysis–based approach and (2) two CNN-based approaches using only SOH and Delta SOH indicators, respectively. Through comparison, the proposed approach demonstrates improved estimation accuracy and robustness. Sensitivity analysis of the CNN and random forest models further validates that the proposed approach makes better use of the available partial discharge data.

The model-based method is similar to SOC estimation, that is, through the establishment of a battery-related model, and then relies on the filtering algorithm and intelligent optimization algorithm to complete the calculation of the battery SOH (Bian et al. 2021; Gholizadeh and Yazdizadeh 2020; S. Hong, Yue, and Liu 2022; M. Lin, Zeng, and Wu 2021). Kalman filter algorithm and its improvement algorithm, particle filtering algorithm, and its improvement algorithm are commonly used to estimate SOH (S. Liu, Dong, et al. 2022; N. Wang, Xia, and Zeng 2023). This kind of method improves the inherent defects of traditional filtering algorithms, which can improve the accuracy of SOH estimation to a certain extent (Ling and Wei 2021; L. Mao, Hu, Chen, et al. 2023; T. Ouyang, Xu, et al. 2022). Mobile average filtering is a commonly used signal processing technology that can smooth the signal curve and remove noise and outliers (Rahimifard et al. 2021; Sakile and Sinha 2021; Xiaodong Zhang, Sun, et al. 2022). Some researchers have improved the traditional mobile average filtering algorithm and proposed a new filtering algorithm that can remove noise and outliers more effectively, thus improving the accuracy of SOH estimation (M. Wu, Wang, and Wu 2023; Z. Wu, Yin, et al. 2022; W. Zhang, Li, et al. 2023). Based on fuzzy control theory and innovatively combining the fuzzy algorithm with EKF, Daniil Fadeev et al. established fuzzy logic to suppress the error caused by measurement noise in the system, and the accuracy of prediction was greatly improved. Jafari et al. (Jafari and Byun 2022) developed a novel PF-based technique for lithium-ion battery RUL estimation, combining a Kalman filter (KF) with a PF to analyze battery operating data. The PF method is used as the core, and extreme gradient boosting (XGBoost) is used for the observation of RUL battery prediction. Due to the powerful nonlinear fitting capabilities, XGBoost is used to map the connection between the retrieved features and the RUL. The life cycle testing aims to gather precise and trustworthy data for RUL prediction. RUL prediction results demonstrate the improved accuracy of our suggested strategy compared to that of other methods. The experiment findings show that the proposed technique can increase the accuracy of RUL prediction when applied to a lithium-ion battery's life cycle dataset. The results demonstrate the benefit of the presented method in achieving a more accurate remaining useful life prediction.

### 1.2.3 Current Status of SOC/SOH Co-Estimation

There is a very close coupling relationship between the two core state parameters SOC and SOH in BMS. By defining the relationship, the two can be connected in series by the actual capacity in the current state (Takyi-Aninakwa et al. 2023; C. Wang, Wang, et al. 2023; Zuolu Wang, Zhao, et al. 2023; Cheng Xu, Zhang, et al. 2022). In the estimation of the two state quantities separately, the time scale in the process of change is different, so the association between the two is usually ignored by researchers (Zuolu Wang, Zhao, et al. 2023; Sheyin Wu, Pan, and Zhu 2022; D. Xiao et al. 2020). However, with the aggravation of the aging degree of lithium-ion batteries, the attenuation of nominal capacity, the change of the relationship curve of battery internal parameters changing with SOC dynamics, and the compression and offset of the SOC-OCV curve will all affect the accuracy of SOC estimation (Xiaoqiang Zhang and Yan 2021). The SOC and SOH co-estimation refers to the estimation of two states, although the relevant functions of BMS bring many repeated calculations (Bian et al. 2022; Saihan Chen, Sun, et al. 2023). The co-estimation of SOC and SOH refers to the estimation of SOH by using the estimation results of SOC, which is very convenient and efficient, that is, reduces redundant calculation and also improves the estimation efficiency (Jo, Jung, and Roh 2021; Kurzweil et al. 2022; Lai et al. 2022). The co-estimation of SOC and SOH is essential for improving the usability of BMS systems. Liu et al. (S. Liu, Dong, et al. 2022) presented an adaptive unscented Kalman filter algorithm (AUKF) for the joint estimation of SOC and SOH of lithium-ion batteries. Firstly, this chapter develops a 2-RC equivalent circuit model and identifies the model parameters using a recursive least squares algorithm with a forgetting factor. Then, the SOC and SOH of the battery are estimated simultaneously by AUKF. Finally, the accuracy of the proposed method is verified under different operating conditions. The experiment results show that the maximum SOC estimation error is under 0.08% in the proposed way. Compared with the unscented Kalman filtering (UKF), the proposed method is more accurate and reliable. A practical approach is provided for state estimation for battery management systems. According to Zhu et al. (F. Zhu and Fu 2021), by integrating the unscented Kalman filter (UKF) and improved unscented particle filter (IUPF) algorithm, SOH can be effectively evaluated. The UKF algorithm is used to estimate the state of charge (SOC), and the IUPF algorithm is employed to identify the ohmic internal resistance. The novelty of the proposed strategy relies on the four-dimensional IUPF filter that is split into a three-dimensional UKF filter and a one-dimensional IUPF filter. Experimental results demonstrate that more accuracy and a faster rate of SOH estimation can be achieved via the UKFIUPF algorithm compared to the IUPF approach.

The method of co-estimation of SOC and SOH can be roughly divided into the following two categories (Yonghong Xu, Chen, et al. 2022; Z. Xu, Wang, et al. 2022; L. Ye, Peng, et al. 2023; S. Zhang, Peng, et al. 2022; T. Zhang, Guo, et al. 2021): The first category is to directly estimate the battery SOH with health indicators and then use the pre-calibrated numerical table to correct the battery impedance and OCV curve and finally realize the estimation of SOC (Obeid et al. 2023; Bavand et al. 2022; C.-Q. Du, Shao, et al. 2022; Gao et al. 2022). The second type is to estimate the SOC and SOH separately, making a real-time comparison of the two estimated state parameters and making the estimation results of the two interact with each other, and feedback correction to achieve the purpose of joint optimization so that both can be closer to the actual value (Guoqi Zhao et al. 2021; J. Zhao, Zhu, et al. 2023; Natella, Onori, and Vasca 2023; S. Park, Kim, and

Cho 2023; P. Ren, Wang, et al. 2022). There is a severe problem with the first type of method being used in practice. That is, it takes a large number of complex experiments and calculations to obtain the corresponding relationship of each quantity with SOH, which takes a long time and is difficult to apply widely (Potrykus et al. 2020; Zuolu Wang, Feng, et al. 2021). The second type of method is relatively feasible, but the calculation details need to be improved to improve the estimation efficiency.

To sum up, particular progress has been made in the co-estimation of SOC and SOH in lithium-ion batteries, and fruitful research results have been accumulated. However, it still needs to be more robust compared with the individual estimation. Combining the two organically and realizing the closed-loop assessment of mutual correction are still the current key research directions.

## 1.3 RESEARCH CONTENT AND FRAMEWORK OF THIS PROJECT

The onboard power lithium-ion battery is the core technical problem of further technological breakthroughs and broad application of new energy vehicles and to realize the real-time accurate estimation of SOC and SOH, which is of great significance to strengthen the real-time monitoring function of power battery management system and ensure the safe and reliable operation of power batteries.

In this book, taking the ternary lithium-ion battery as the research object, the battery charge and discharge experiments under complex working conditions are analyzed, and a dynamic battery migration model that can accurately describe the characteristics of the battery is established. The double filter is constructed, and the population intelligent optimization of the firefly algorithm is introduced to achieve a powerful high-precision SOC and SOH collaborative estimation of lithium-ion batteries. It effectively provides real-time and accurate residual power status and health status, extending the service life of power lithium-ion batteries and providing a solid guarantee for the safety of electric vehicles.

The main research contents and chapters of this subject are divided as follows:

The first chapter introduces the background and significance of this research topic in detail and presents the current status of SOC estimation, SOH estimation, and SOC/SOH co-estimation. Finally, the main research contents and the overall research framework of this topic are proposed according to the deficiencies in the current research.

The second chapter is the experimental characteristics analysis of the power lithium-ion battery. After the current and voltage data under three complex working conditions are obtained by formulating and conducting experiments, the functional relationship between the open circuit voltage and SOC is obtained by analyzing and extracting the data and curve fitting. The charge and discharge characteristics and capacity attenuation characteristics are studied, the influence of discharge rate on SOC change and the influence of capacity attenuation on battery internal parameters and SOH are analyzed, and the relationship between aging characteristics and SOH change is studied. It provides the basis for the lithium-ion battery modeling part in Chapter 3.

The third chapter is the dynamic migration model construction of a power lithium-ion battery. First, the initial Thevenin model is established, and the functional relationship between each parameter and the battery SOC is obtained using offline parameter identification. The migration factor is added to the functional relationship to realize the adjustment of each parameter. The migration factor forms the migration matrix, and the PF algorithm is used to continuously correct and update the migration matrix to realize the dynamic adjustment of the battery parameters.

The fourth chapter is a co-estimation study of SOC and SOH for power lithium-ion batteries based on the firefly optimization algorithm. This chapter discusses the progressive reciprocal correction of SOC and SOH. The first layer particle filter realizes the SOC estimation, and the estimated value is used as the input of the second layer filter to learn the progressive estimation of SOH. Then, the current moment SOH estimate obtained in the second-layer Kalman filter is taken as input for the next iteration cycle, and the SOC estimate of the next moment is further corrected to form a closed loop until the end of the iteration. Then, the firefly algorithm is applied in the optimization process of SOC and SOH particles to realize the proximity of the particle to the optimal value, that is, to achieve the purpose of further proximity to the actual value. On this basis, the chaos mapping algorithm is added, which linearly maps the variables into chaotic variables through chaotic mapping. Then, it optimizes the search process according to the ergodicity and randomness of the chaos, and the high precision co-estimation of SOC and SOH is realized.

Chapter 5 is the experimental verification and analysis of the cooperative estimation of SOC and SOH. In this chapter, to verify the effectiveness of the proposed dynamic migration model and the population optimization firefly algorithm, the co-estimation of SOC and SOH under high health conditions is discussed. To verify the adaptive regulation ability of the proposed dynamic migration model for different aging conditions, the co-estimation of SOC and SOH was verified and analyzed under mild aging and severe aging, respectively, further ascertaining the effectiveness of the model and algorithm.

Chapter 6 is the summary and prospect of the work on this topic. The work and the results of the institute are summarized, the shortcomings in the current research methods are analyzed, and the subsequent research work of this book is discussed.

CHAPTER 2

# Working Principle and Experimental Characteristic Analysis of Power Lithium-Ion Battery

THE POWER LITHIUM-ION BATTERY is one of the primary power sources of new energy vehicles, which has very complex dynamic characteristics in practical application. This chapter introduces the structure and working principle of the lithium-ion battery and analyzes the internal operation mechanism of the battery in the working process and the primary operating characteristics of the lithium-ion battery, laying a foundation for the following.

## 2.1 WORKING PRINCIPLE ANALYSIS OF POWER LITHIUM-ION BATTERY

The ternary lithium-ion battery has been widely used in new energy vehicles because of its advantages of high energy density, low self-discharge, low pollution, long cycle life, and excellent high-temperature and low-temperature resistance (Cai et al. 2023; H. Ji et al. 2020; Lei Liu, Lin, et al. 2020; Yan Liu, Lv, et al. 2021). In this part, we will conduct an in-depth analysis and research


DOI: 10.1201/9781003486251-2

on the ternary lithium-ion batteries used for automobiles. The ternary lithium-ion battery is composed of the positive electrode, negative electrode, electrolyte, and isolation layer, and the positive electrode is formed of lithium-ion cobalt oxide, lithium-ion iron phosphate, Ni-Co-Al or Ni-Co-Mn ternary substance (G. Zhang, Li, et al. 2022; J. Zhang, Jin, et al. 2021; X. Bai, Ban, and Zhuang 2020; Guan et al. 2023; Z. Hu, Huang, et al. 2023). The cathode is mainly graphite. Between the two extremes of the battery, filled with electrolyte, it acts as a medium and bridge so that lithium-ion can move between the two extremes (Guange Wang, Zhang, et al. 2020; X. Wu, Tang, et al. 2023; C. Yang, Zhang, et al. 2022). The diaphragm mainly plays the role of separating the positive and negative electrodes of the lithium-ion battery, an organic film with micropores, like the baffle between the positive and negative electrodes. It only allows lithium-ions to pass through (K. Huang, Xiong, et al. 2022; Yue Li, Huang, Li, et al. 2023; Yongkun Li, Wei, et al. 2020; Zongwei Liu, Liu, et al. 2020). The internal working principle and operation mechanism of lithium-ion batteries are shown in Figure 2.1.

The chemical reaction process inside the lithium-ion battery mainly covers the charging chemical reaction and the discharge chemical reaction (D. Ren et al. 2021; Stinson 2023). As can be seen from Figure 2.1, during

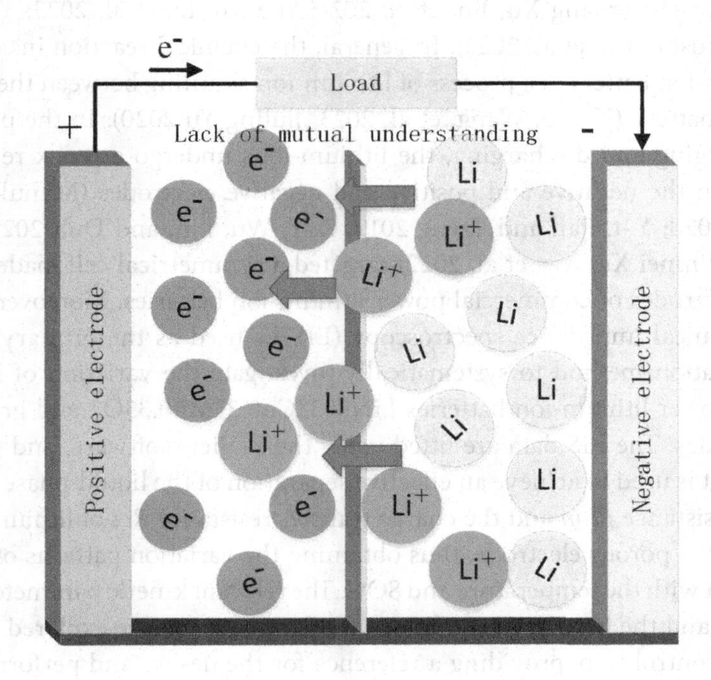


FIGURE 2.1 Working principle of lithium-ion battery.

the chemical reaction of charging, lithium-ion is accessible from the positive side (usually oxide), and freely moving ions travel through the medium of the electrolyte into the pores of the cathode material (usually graphite). At the same time, the external power supply of the battery takes the electrons from the cathode material. It enters the cathode material through the external circuit so that the cathode material is oxidized (Zhuang et al. 2020). In this process, the concentration of lithium-ions in the electrolyte inside the battery gradually decreases. In contrast, the attention in the cathode material gradually increases until the battery is fully charged (G.H. Chang et al. 2020; Jiangtian Sun et al. 2021; Junlian Wang, Fu, et al. 2020; Y. Xie et al. 2021). During discharge, the external load causes electrons to flow out of the cathode material and enter the cathode material through the external circuit, causing the cathode material to be reduced (Dai et al. 2023; Y. Ge et al. 2023; L. Jiang et al. 2019; L'Vov, Tikhonchev, and Sibatov 2022). At the same time, lithium-ion leaves the cathode material and passes through the electrolyte into the micropores of the cathode material so that the cathode material is oxidized. In this process, the concentration of lithium ions in the electrolyte inside the battery gradually increases, while the attention in the cathode material gradually decreases until the battery runs out (Jianguang Xu, Jin, et al. 2021; Yue Xu, Li, et al. 2023; Yoneda 2023; Yoshimura et al. 2022). In general, the chemical reaction inside the lithium-ion battery is a process of lithium-ion shuttling between the poles of the battery (Y. Xie, Wang, et al. 2023; Jiuling Yu 2020). In the process of charging and discharging, the lithium-ions undergo a redox reaction between the negative and positive and negative electrodes (Manukumar et al. 2020; Y.-t. Pan and Tzeng 2019; C.-Y. Wu, Lin, and Duh 2022). Xu et al. (Jinmei Xu, Xie, et al. 2022) targeted a symmetrical cell made from the electrodes of commercial power lithium-ion batteries. Moreover, electrochemical impedance spectroscopy (EIS) is used as the primary characterization method to systematically investigate the variation of EIS of high-power lithium-ion batteries LiNi0.33Co0.33Mn0.33O2 and graphite electrodes. The EIS data are fitted using the Z-View software, and the Ls element is used to achieve an effective separation of the liquid-phase diffusion resistance *Rion* and the charge transfer resistance *Rct* of lithium ions inside the porous electrode, thus obtaining the variation patterns of Rion and Rct with the temperature and SOC. The relevant kinetic parameters are tested, and the boundary conditions for Rion and Rct are explored as the speed control step, providing a reference for the design and performance improvement of commercial high-power lithium-ion batteries, especially those in specific temperature and SOC intervals. In Jinmei Xu, Yang, et al.

(2023), the authors explored the changing tendencies of the Li+ diffusion resistance R-ion and charge transfer resistance R-ct with temperature and SOC. Moreover, one quantitative parameter is introduced to identify the boundary conditions of the rate control step inside the porous electrode. This work points out the direction to design and improve performance for commercial HEP LIB with a standard temperature and charging range of users. Through the analysis of the internal working principle of the battery, we can make sufficient preparation for the subsequent study of the working characteristics of the battery.

## 2.2 EXPERIMENTAL DESIGN AND CONDUCT OF POWER LITHIUM-ION BATTERY

### 2.2.1 Construction of the Experimental Platform

In this study, the nickel-cobalt-manganese ternary lithium-ion batteries with a rated capacity of 45 Ah, a nominal voltage of 3.7 V, discharge and charging cut-off voltages of 2.75 V and 4.2 V were selected as the research objects, and the BTS200–100–104 battery test equipment was used as the test platform to build the experimental platform. To avoid the effects of temperature discordance on cell parameters and OCV, all experiments performed in this study were performed at a constant temperature of 25°C. The constructed experimental platform is shown in Figure 2.2.

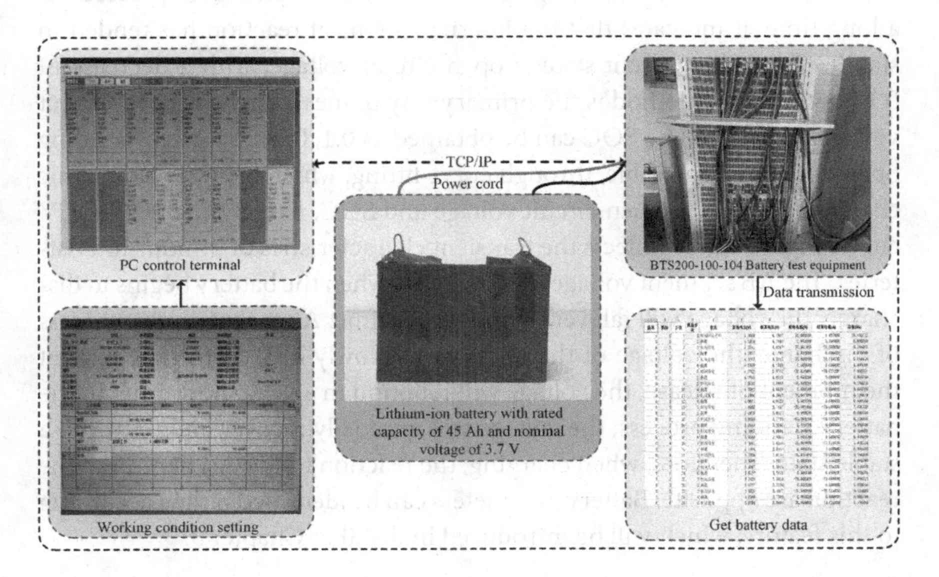


FIGURE 2.2 Experimental platform of lithium-ion battery.

### 2.2.2 Hybrid Pulse Power Characterization Experiment

Hybrid pulse power characterization (HPPC) is a basic experiment of lithium-ion batteries which is a standardized test method for the performance evaluation of electric vehicle batteries. Its primary purpose is to determine the capacity and internal resistance of the battery, which is often used as the experimental verification condition of the model and algorithm. In the HPPC experiment, the cell is applied to a series of pulse currents to analyze the cell performance by measuring the voltage response of the cell. The operation steps of the HPPC test are as follows: first, charge a lithium-ion battery with unknown current remaining power condition to the cut-off voltage, set it for 1 h after filling, then discharge for 1 h, then discharge the lithium-ion battery for 10 s, charge for 40 s, then rest for 40 s. The whole process is charged and discharged for 1 C. This step is recycled ten times with an hour experimental interval between adjacent pulses (the battery is discharged at 1 C current for 6 min, and the SOC value decreased by 0.1). In the HPPC experimental cycle test process, the whole process voltage response curve and the single voltage response curve (SOC = 0.7) are shown in Figure 2.3.

From Figure 2.3(a), it can be seen that after each constant current discharge of the lithium-ion battery, there is a significant change in the measured voltage after 1 h of release. Through this phenomenon, it can be analyzed that the battery voltage gradually stabilizes after being placed for a long time. It indicates that the internal chemical reaction has tended to stabilize, and the current state is open circuit voltage, with a decrease of 0.1 in SOC. This method is the primary way of measuring the open circuit voltage. Therefore, the SOC can be obtained as 0.1, 0.2 . . . 1 corresponding open circuit voltage value, through curve fitting, which brings the dynamic change between the open circuit voltage and SOC, namely, the SOC-OCV curve. Figure 2.2(b) reflects the transient characteristics of lithium-ion batteries. The AB segment voltage indicates that when the battery begins to discharge, the voltage will fall very rapidly over time. After that, in the process of discharge, the voltage of the battery will slowly drop with time, and at the moment of release, the voltage will rebound in a very short time. In the battery standing process, the voltage will gradually recover and return to a stable state. Therefore, when charging, the reaction to the discharge and the reaction are opposite. Battery parameters can be identified offline according to this feature, which will be introduced in detail in Chapter 3.

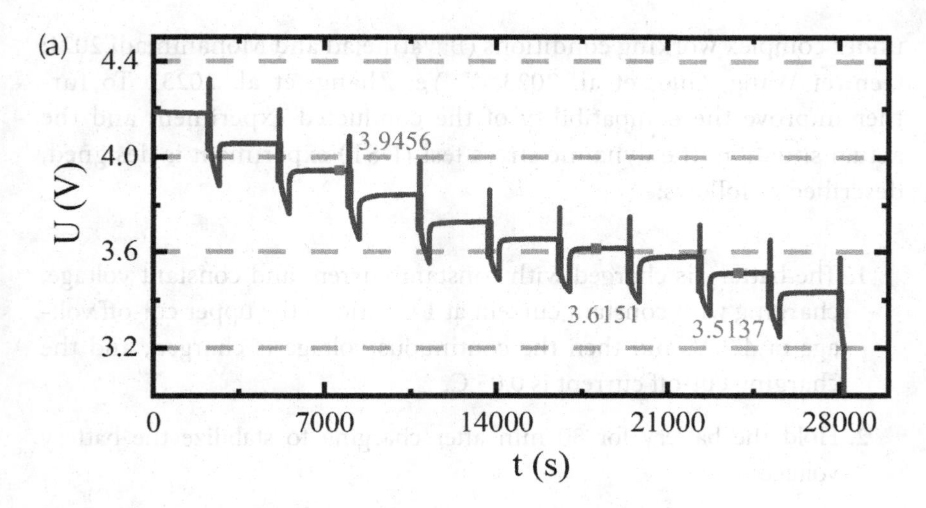


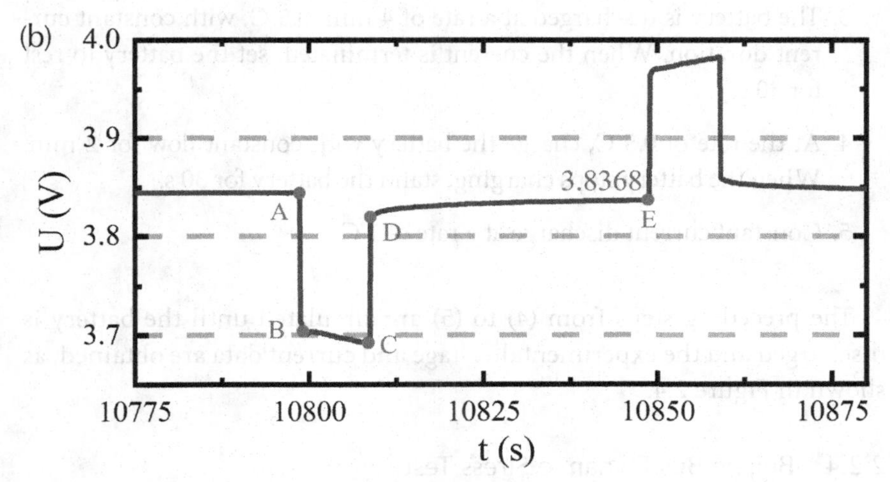


FIGURE 2.3 Experimental data of the HPPC working condition: (a) voltage response curve for the HPPC test; (b) voltage response curve at SOC = 0.7.

### 2.2.3 Dynamic Stress Test Experiment

In the practical application of new energy vehicles, the current is complex and changeable. In different working conditions, it is often accompanied by sudden switching and stopping of current, which puts forward strict requirements for the dynamic performance of batteries. It also brings difficulties to the estimation of SOC and SOH of lithium-ion batteries under complex working conditions (Bayatinejad and Mohammadi 2021; Genwei Wang, Guo, et al. 2023; Y. Ye, Zhang, et al. 2023). To further improve the compatibility of the conducted experiment and the actual situation, the dynamic stress test (DST) experiment is designed, described as follows:

1. The battery is charged with constant current and constant voltage, charging with constant current at 1 C ratio to the upper cut-off voltage of 4.2 V, and then the continuous voltage is charged, and the charging cut-off current is 0.05 C.
2. Hold the battery for 30 min after charging to stabilize the battery voltage.
3. The battery is discharged at a rate of 4 min, 0.5 C, with constant current duration. When the current is terminated, set the battery to rest for 30 s.
4. At the rate of 0.5 C, charge the battery with constant flow for 2 min. When the battery stops charging, stand the battery for 30 s.
5. Constant current discharge at a rate of 1 C.

The preceding steps from (4) to (5) are circulated until the battery is discharged and the experimental voltage and current data are obtained, as shown in Figure 2.4.

### 2.2.4 Beijing Bus Dynamic Stress Test

To better simulate the dynamic condition, variability, and irregularity in the actual operation of new energy vehicles, the Beijing bus dynamic stress test (BBDST) was designed, and the experiment was carried out with the ternary lithium-ion battery as the experiment object. BBDST is the working condition data collected from the actual operation of Beijing city buses, including not only the primary working conditions, such as starting, braking, and parking, but also the acceleration, sliding, and rapid acceleration, which is closer to the actual operation situation of new energy vehicles and lays a foundation for the subsequent better verification of the model and algorithm. The experimental voltage and current data were obtained as shown in Figure 2.5.

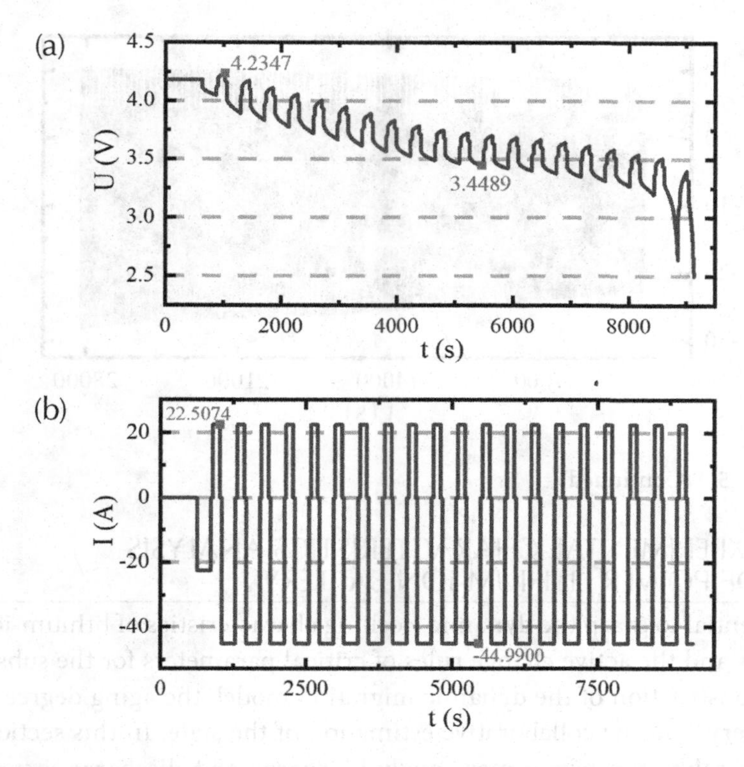


FIGURE 2.4 Experimental data of the DST operating conditions: (a) DST operating condition voltage data; (b) DST operating condition current data.

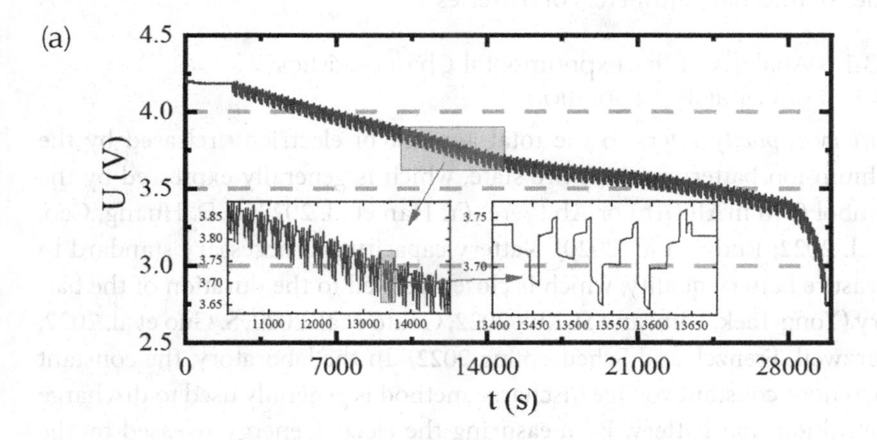


FIGURE 2.5 Experimental data of BB DST working condition: (a) operating condition voltage data of BBDST; (b) operating condition current data of BBDST.

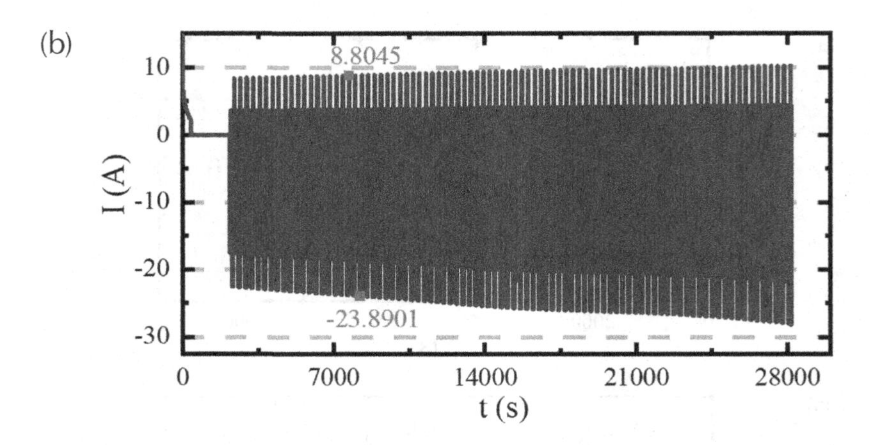


FIGURE 2.5 (Continued)

## 2.3 EXPERIMENTAL CHARACTERISTICS ANALYSIS OF POWER LITHIUM-ION BATTERY

It is essential to grasp the dynamic working characteristics of lithium-ion batteries and the active change rules of critical parameters for the subsequent construction of the dynamic migration model, the aging degree of the battery, and the collaborative estimation of the state. In this section, capacity calibration experiment analysis, charge and discharge experiment analysis of different ratios, and battery aging test analysis are carried out in a 25°C constant temperature environment to explore the change rules of internal parameters of batteries.

### 2.3.1 Analysis of the Experimental Characteristics of Capacity Calibration

*Battery capacity* refers to the total amount of electricity released by the lithium-ion battery in its entire state, which is generally expressed by the symbol $Q$ in mAh (Ah) or Ah (pere) (Y. Han et al. 2023; K.D. Huang, Cao, et al. 2022; Rudyi et al. 2020). Battery capacity is a necessary standard to measure battery quality, which is closely related to the duration of the battery (Yong-Taek, Kim, and Si-Kuk 2022; Couto et al. 2022; S. Guo et al. 2022; Kurzweil, Frenzel, and Scheuerpflug 2022). In the laboratory, the constant current or constant voltage discharge method is generally used to discharge the lithium-ion battery. By measuring the electric energy released by the battery during the whole discharge process or calculating the product of the discharge current and time, this is the storage capacity value of the lithium-ion battery (Junfu Li, Xu, et al. 2022; N. Li, He, et al. 2022; Ran Li, Wei, et al.

2022; Lv, Huang, and Liu 2020; Makeen, Ghali, and Memon 2020). After the continuous use of lithium-ion batteries, the actual capacity of the new battery is certainly less than the actual capacity of the new battery.

Due to the aging of the recycling battery, the battery capacity will have a significant deviation from the calibration capacity of the factory. The actual discharge capacity of the battery is of great significance for the estimation of SOC and SOH of the lithium-ion battery (Kumar and Pareek 2023; Lihua Liu, Zhu, and Zheng 2020; You et al. 2021). Therefore, the capacity calibration experiment of the lithium-ion battery at a 25°C constant temperature environment was conducted first. The specific steps are described as follows:

1. The battery is charged at a 1 C rate to a charging voltage of 4.2 V and then converted to a constant voltage charging current of 0.02 C to ensure that the battery is in full charge state.
2. Let the battery sit for 40 min.
3. The constant discharge test was carried out on the lithium-ion battery at a 1 C rate, and the discharge cut-off voltage was 2.75 V.
4. Let the battery sit for 20 min.

Cycle steps (1)–(4) three times to end the experiment. The changes in battery capacity throughout the capacity calibration experiment are shown in Figure 2.6.

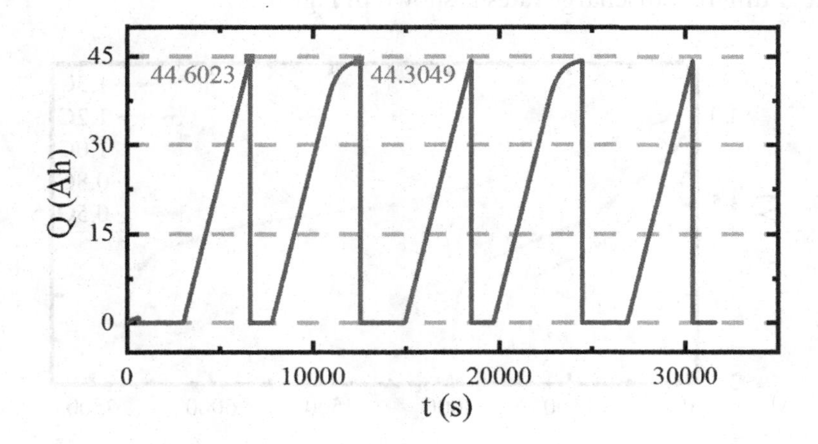


FIGURE 2.6 Capacity calibration experiments.

The tertiary discharge capacity is averaged, and the actual capacity of the battery in the current state is 44.6 Ah. Compared with the rated capacity of 45 Ah, the lithium-ion battery is slightly aging at this time, and the capacity has a small decay. Capacity calibration experiments are the basis for subsequent co-estimation of SOC and SOH.

### 2.3.2 Analysis of Charge and Discharge Characteristics of Different Ratios

The *charge–discharge rate* of a battery refers to the ratio of the current of the battery during charging and discharging to its rated capacity. Usually, with *C*, such as 1 C charge or 1 C discharge, it means that the charge or discharge current of the battery is twice its rated capacity. For example, if the capacity of a battery is 1,000 mAh, its 1 C charging current and 1 C discharge current is 1,000 mA. If the charging current is 500 mA, then its charging rate is 0.5 C, and if the discharge current is 2,000 mA, then its discharge rate is 2 C. The charge and discharge rates of the battery are of great significance for its use. If the charge and/or discharge rate is too high, it will lead to problems such as battery heating and capacity decline, which will affect battery life. Therefore, it is necessary to select the appropriate charge and discharge rates according to the specifications and design requirements of the battery to ensure the safety and reliability of the battery. In this section, the influence of discharge rate on the battery was studied qualitatively. Different discharge experiments ($I$ = 1.5 C, 1.2 C, 1.0 C, 0.8 C, 0.5 C) were conducted on the battery, and the monitored output voltage value was sampled one by one at 0.01 s. The end voltage curve under different discharge rates is shown in Figure 2.7.

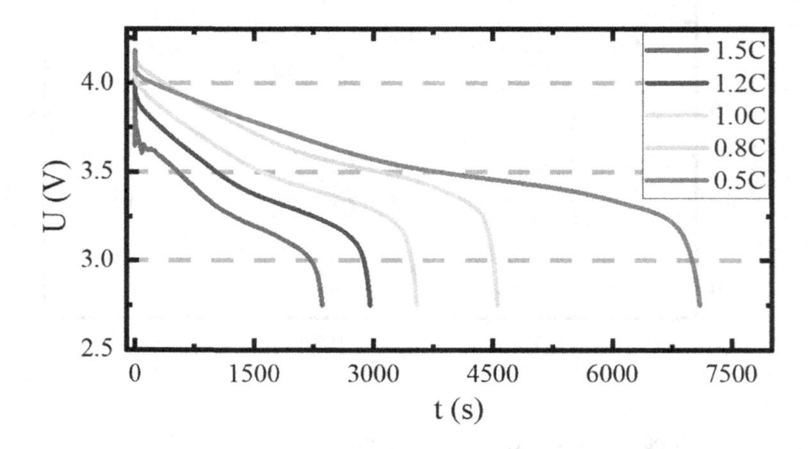


FIGURE 2.7 Charge and discharge experiments with different ratios.

By observing the voltage change curve of the battery end at different discharge rates in Figure 2.7, it can be concluded that the change of the end voltage is extremely nonlinear during the whole constant current discharge process. The battery discharge process can be divided into three stages. In the first stage, the voltage will transiently decrease with the current change, which is due to the instantaneous ohm polarization inside the battery, and the ohm internal resistance is divided. In the second stage, the voltage of the battery stabilizes because the electrochemical reaction state inside the battery is flat, and the voltage difference between the positive and negative electrodes slowly decreases. In the third stage, when the battery discharges near 3.2 V, the end voltage curve will change and almost show a straight-line decline. By comparing the five different voltage change curves in Figure 2.7, we can find that with the increase of the battery discharge rate, the total discharge time will decrease successively. In the voltage curve, the "plateau" of the voltage becomes more and more "steeper," which will lead to the actual insufficient discharge. But no matter what rate the battery discharges, the total discharge capacity is the same.

### 2.3.3 Analysis of Battery Aging Test Characteristics

The aging of a battery refers to the phenomenon that the performance and life of the battery gradually decrease with the passage of time and the use of the battery (Boerger et al. 2022; X. Chang, Zhao, et al. 2023; Galos et al. 2021). The joint action of various reasons causes the phenomenon of battery aging. The chemical reaction inside the battery in the work is the basis for the battery to store and release energy (Heimes et al. 2020). In the process of charging and discharging, the positive and negative electrode materials inside the battery will undergo a REDOX reaction (Heins et al. 2020; W. Hu and Zhao 2022; K.D. Huang, Cao, et al. 2022; R. Huang et al. 2023), which will lead to the constant change of the chemicals inside the battery and produce various reaction products. These reaction products will gradually accumulate on the surface of the electrode and the electrolyte, forming the dirt and membrane layer inside the battery, which will hinder the ion transmission and charge transmission inside the battery, thus affecting the performance and life of the battery (Ibrahim et al. 2023; Xinyu Jia, Zhang, et al. 2022; J. Lee et al. 2021). In addition, the internal resistance of the battery will gradually increase with the increase of the charge and discharge cycle. This is because the intense chemical reaction of the battery during the operation causes changes in the structure and material properties of the battery, increasing the resistance inside the battery. At the same time, the temperature and use mode will also affect the internal resistance of the battery (J. Liu, Duan, et al. 2022; Menz et al. 2023; Osara et al. 2021; S.Y. Park et al. 2021; Redondo-Iglesias, Venet, and Pelissier 2020). In short, the aging of the battery is caused by the reaction products of the internal chemical reaction of the battery, dirt, and membrane layer, increased internal resistance, and other factors. To extend the service life of the battery, it is necessary to take appropriate charging and discharging methods to avoid excessive charging and discharging and high-temperature use (J. Shi, Rivera, and Wu 2022; Willenberg et al. 2020). At the same time, it is also necessary to maintain the battery regularly and remove the dirt and membrane layer inside the battery to maintain the performance and life of the battery.

Battery capacity is an important indicator of the battery energy storage capacity, and its size is positively correlated with the health degree of the battery (Dechent et al. 2021; Ould Ely, Kamzabek, and Chakraborty 2019). Therefore, by measuring the capacity value of the battery, we can achieve the purpose of assessing the health status of the battery (M. Kim and Han 2021; Levieux-Souid et al. 2022; Vahnstiege et al. 2023). It is essential to study the relationship between the battery aging phenomenon and cycle times. In this study, the accelerated aging experiment on lithium-ion batteries, accelerated cycle charge and discharge 500 times, measured and recorded the discharge power of the battery during each cycle. The change of capacity $Q$ with the increase of battery cycles is shown in Figure 2.8.

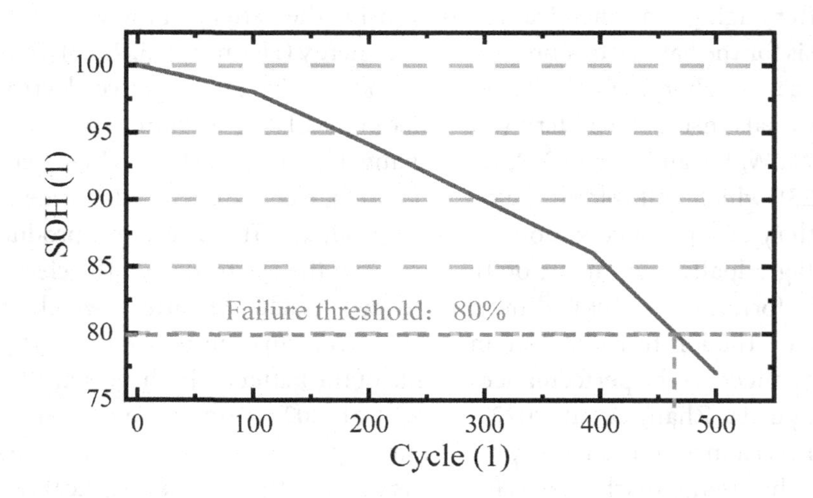


FIGURE 2.8 Battery aging test.

In Figure 2.8, the *cycle* is the number of cycles. As can be seen from the figure, with the increase of the charge and discharge times of the battery cycle, the battery SOH gradually decreases. When the cycle reaches 466 times, the battery capacity drops to the failure threshold, and the decline speed of the SOH after this point is accelerated. This shows that the capacity of the battery changes during use rather than maintaining a fixed value. Therefore, it is necessary to consider the influence of the capacity decline phenomenon in the process of SOC estimation.

## 2.4 SUMMARY OF THIS CHAPTER

In this chapter, the author first explored and analyzed the internal structure and working principle of the onboard power lithium-ion battery in new energy vehicles, completed the configuration and construction of the experimental platform of the system, formulated the lithium-ion battery experiments under three complex working conditions, and completed the practical development and data sorting. At the same time, the experimental characteristics of lithium-ion battery capacity calibration were analyzed to measure the actual capacity of the selected battery under the current state, examine the charge and discharge characteristics of different ratios, explore the influence of various sizes on the voltage change of the battery end, explore the dynamic correlation between cycle times and capacity attenuation, and analyze the change mechanism. The research content of this chapter lays the foundation for the subsequent construction of a dynamic migration model and synergistic state estimation considering aging characteristics.

CHAPTER 3

# Dynamic Migration Modeling of Power Lithium-Ion Batteries

Battery aging is an essential factor affecting the regular use of onboard power lithium-ion batteries in the operation of new energy vehicles. The uncertain aspect of battery aging dramatically affects the accuracy of battery modeling. Therefore, in this chapter, battery migration modeling is introduced, and the impact of battery aging is regarded as an uncertain dynamic factor. The initial Devinin model is established by identifying the relationship between offline parameters and battery charging state function. Then, the particle filter algorithm is used to modify and update the parameters to achieve dynamic adjustment, and the optimal parameters of a set of functional relationships can be found at each sampling point. It dramatically improves the accuracy of the battery model and lays a solid foundation for the subsequent collaborative estimation of SOC and SOH.

## 3.1 DYNAMIC MIGRATION MODELING

### 3.1.1 Offline Initial Model Establishment

The battery equivalent circuit model is a method to simulate the battery's internal chemical reaction process with the circuit elements, which can describe the electrical properties and characteristics of the battery



DOI: 10.1201/9781003486251-3

(Z. Geng et al. 2021; Sato et al. 2019; Song, Peng, and Liu 2021). Usually, the battery equivalent circuit model is composed of the battery potential, internal resistance, capacitance, and other components (S. Wang, Cao, et al. 2022). Among them, the *battery potential* refers to the potential difference between the positive and negative electrodes of the battery, which is the basis on which the battery can generate electricity. In the cell equivalent circuit model, the cell potential is usually expressed as a fixed potential source whose size is equal to the nominal voltage of the cell (Yujie Wang and Zhao 2023; D. Yu, Ren, et al. 2021). The chemical reaction inside the battery leads to an increase in the resistance inside the battery, which affects the output current of the battery and the maintenance time of the battery. In the battery equivalent circuit model, the internal resistance is usually expressed as a resistance element, which is connected in parallel with the battery potential (Yubai Li, Zhou, and Wu 2020; Su et al. 2019; J. Xie, Wei, et al. 2023). The chemical reaction process inside the battery also leads to the capacitance change inside the battery, which affects the transient response of the battery and the stability of the output current. In a cell equivalent circuit model, a capacitor is usually represented as a capacitive element that is connected in series with the battery potential (Brucker et al. 2022; Karimi et al. 2022; Q. Wang, Gao, and Li 2022). To sum up, the battery equivalent circuit model is usually composed of components such as battery potential, internal resistance, capacitor, and on, which can simulate the internal chemical reaction process of the battery and describe the electrical performance and characteristics of the battery (B. Wang, Qin, et al. 2020). In practical applications, the battery equivalent circuit model can be used to predict parameters such as the battery output current, voltage, and maintenance time to evaluate the performance and life of the battery (Drees, Lienesch, and Kurrat 2022; Hao Li, Zhang, et al. 2022; Pang et al. 2021). The standard equivalent circuit model includes the internal resistance comparable circuit model, Thevenin equivalent circuit model, second-order RC equivalent circuit, PNGV equivalent circuit model, etc.

The Thevenin model (M.U. Ali, Khan, et al. 2022; Bobobee et al. 2023; D. Chen, Xiao, et al. 2021) is one of the widely used equivalent circuit models of lithium-ion batteries. Compared with the traditional internal resistance model, its improvement lies in the addition of an RC circuit to characterize the polarization effect in the working case of lithium-ion battery (Shan Chen, Pan, and Jin 2023; He et al. 2021; W. Ji et al. 2021; Xintian Liu, Deng, et al. 2020). Thevenin model can better characterize the dynamic response of the battery and the battery internal resistance $R_0$

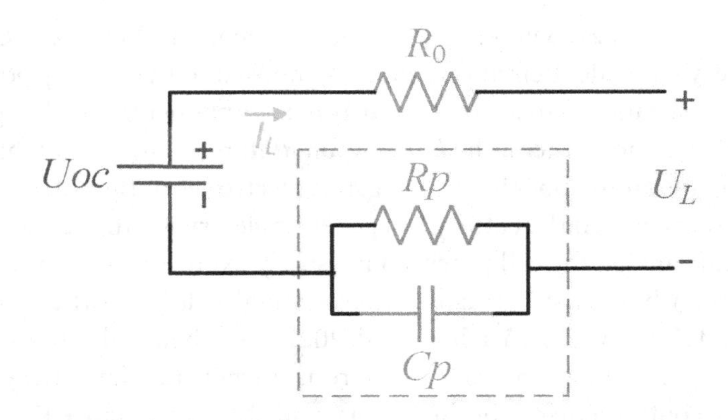


FIGURE 3.1 The offline initial Thevenin model.

to represent the instantaneous voltage response of battery charge and discharge (Yuyang Liu, Wang, Xie, et al. 2022). The RC circuit is a polarized internal resistance $R_P$ and the polarized capacitance $C_P$. The composed circuit can reflect the gradual change of the battery voltage during and after charging and discharge. The Thevenin equivalent circuit model is not only simple in structure but also can meet the simulation requirements (Qi et al. 2022; K. Wang, Feng, et al. 2020; M. Wu, Qin, and Wu 2021; W. Xu, Wang, et al. 2020). Considering the model accuracy and calculation quantity, the Thevenin equivalent circuit model is established as the offline initial model, as shown in Figure 3.1.

### 3.1.2 Identification of Model Parameters and Extraction of Relationship Curves

The establishment of a dynamic migration model requires obtaining a function between the initial battery model parameters and SOC rather than the specific parameter value corresponding to each sampling point under a particular condition (Takyi-Aninakwa et al. 2023; Bayatinejad and Mohammadi 2021; J.-J. Li, Dai, and Zheng 2021). Therefore, online parameter identification in migration modeling is not applicable (Junhao Qiao et al. 2023; Smith and Siegel 2020; X. Wang and Tong 2023). With Thevenin parameter $R_0$ as included in the model, $R_P$ and $C_P$ can be identified offline by the processing and analysis of the HPPC experimental data obtained in Chapter 2.

In the process of data analysis and processing, the effective data segments are extracted from the original experimental data. Through the analysis of the extracted data segments and the processing method of curve fitting, the relationship between internal parameters and SOC is obtained, and the accurate construction of the equivalent model is completed to realize the accurate description of the working characteristics of lithium-ion batteries (C. Chang, Wang, et al. 2023; Pai, Liu, and Ye 2023; L. Xu, Pan, et al. 2022). First, all voltage data were extracted from the raw data to describe the battery end voltage changes throughout the HPPC test, as shown in Figure 3.2(a). With parameter $R_0$ to be identified for the initial Thevenin model, $R_P$, and $C_P$, the pulse test data segment corresponding to each cycle in the HPPC test is extracted. Taking SOC = 0.7 as an example, the pulse response curve of a lithium-ion battery is shown in Figure 3.2(b).

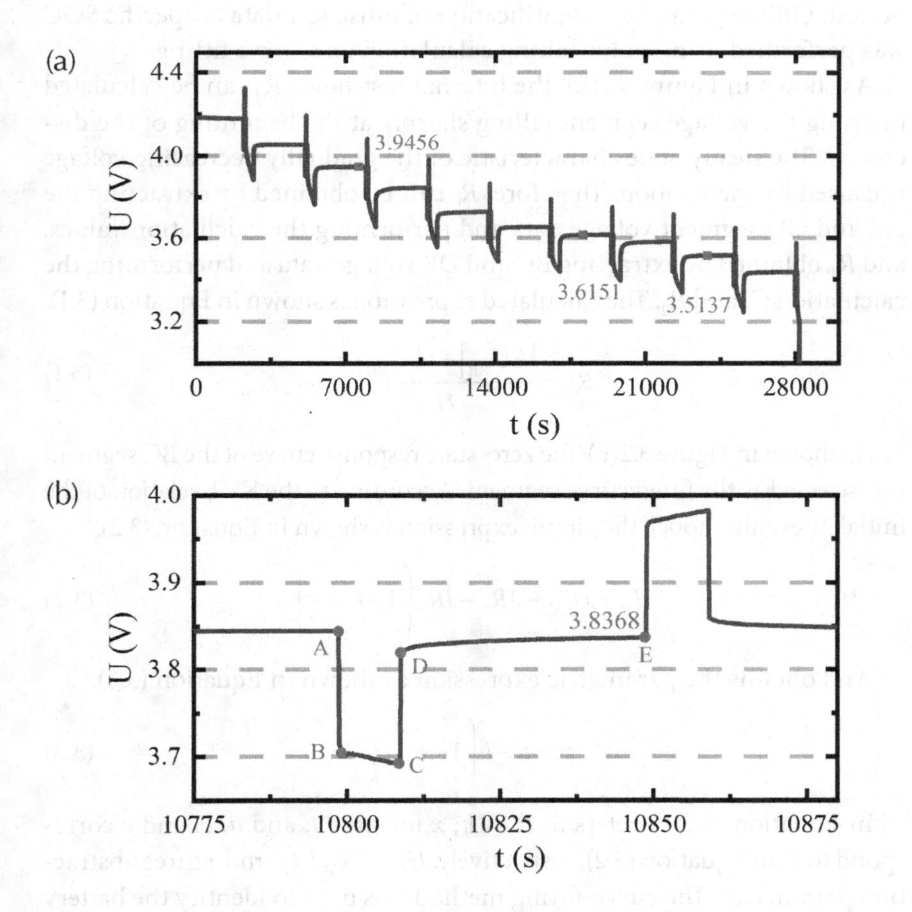


FIGURE 3.2 Voltage response curve of HPPC working condition: (a) voltage response curve of the HPPC test; (b) voltage response curve at SOC = 0.7.

As can be seen from Figure 3.2(a), each time the battery discharges through constant current and the battery is shelved for 1 h, the voltage gradually stabilizes, which indicates that the electrochemical reaction and thermal effect inside the battery have reached balance. At this point, the measured battery voltage is its open circuit voltage so that you can measure the SOC of 0.1, 0.2, etc. The open circuit voltage at 1 follows the curve between OCV and SOC. Figure 3.2(b) reflects the transient and steady-state characteristics of lithium-ion batteries. At the beginning of the pulse discharge, the battery voltage changes and drops instantly, and the voltage slowly drops during the constant current discharge. At the end of the release, the battery voltage rebounds immediately, and then the voltage slowly rises and gradually tends to stabilize during the shelved period. Offline parameter identification of pulse test data at specific SOC was performed using point-taking calculation and curve fitting.

As shown in Figure 3.2(b), the internal resistance $R_0$ can be calculated by using the voltage segment falling sharply at the beginning of the discharge. The steady-state characteristic of the gradually decreasing voltage is caused by the $R_C$ loop. Therefore, $R_0$ can be obtained by extracting the AB and CD segment voltage data and performing the calculation values, and $R_P$ obtained by extracting BC and DE voltage data and performing the calculation $C_P$ and $R_0$. The calculated expression is shown in Equation (3.1).

$$R_0 = \frac{|\Delta U_{AB}| + |\Delta U_{CD}|}{2I} \tag{3.1}$$

As shown in Figure 3.2(b), the zero-state response curve of the BC segment was selected as the fitted curve segment. According to the KVL relation of the initial Thevenin model, the circuit expression is shown in Equation (3.2).

$$U_L = U_{OC} - IR_0 - IR_p\left(1 - e^{-\frac{\Delta T}{\tau}}\right) \tag{3.2}$$

And obtains the parametric expression as shown in Equation (3.3).

$$y = a - b\left(1 - e^{-\frac{x}{c}}\right) \tag{3.3}$$

In Equation (3.2), $y$ refers to the $U_L$; $x$ for time $t$; and $a$, $b$, and $c$ correspond to $U$ in Equation (3.2), respectively, $U_{OC}-IR_0$, $IR_P$, and $\tau$ three abstraction parameters. The curve-fitting method was used to identify the battery parameters for each stage with an SOC of 0.1 to 1, as obtained in $R_0$, $R_P$, and $C_P$. The offline parameter identification results are shown in Table 3.1.

TABLE 3.1 Offline Parameter Identification Results

| SOC | $R_0$ /Ω | $R_p$ /Ω | $C_p$ /μF |
|---|---|---|---|
| 1 | 0.0028 | 0.0005752 | 14,361.9611 |
| 0.9 | 0.0032 | 0.0006374 | 11,909.3191 |
| 0.8 | 0.0028 | 0.0006786 | 11,497.2001 |
| 0.7 | 0.0024 | 0.0007104 | 11,392.1734 |
| 0.6 | 0.0032 | 0.0006286 | 11,800.8272 |
| 0.5 | 0.0013 | 0.0004946 | 17,270.5216 |
| 0.4 | 0.0022 | 0.0004876 | 17,518.4578 |
| 0.3 | 0.0037 | 0.0005244 | 17,080.4729 |
| 0.2 | 0.0034 | 0.000601 | 15,094.8419 |
| 0.1 | 0.0029 | 0.0008518 | 9,839.16412 |

$R_0$ was obtained using SOC as the independent variable and each battery parameter as the dependent variable, $R_P$, $C_P$ scatter plot of the relationship between an SOC, and fitted curves obtained by least squares polynomial fitting method. The expression of the function describing the dynamic relationship between three battery parameters and SOC is shown in Equation (3.4).

$$\begin{cases} R_0 = -0.02978x^6 + 0.09091x^5 - 0.105x^4 + 0.05578x^3 - 0.01196x^2 \\ \quad + 0.0001415x + 0.003219 \\ R_p = 0.06207x^6 - 0.189x^5 + 0.2216x^4 - 0.1317x^3 + 0.04575x^2 \\ \quad - 0.009677x + 0.001471 \\ C_p = -1.165e{+}07x^6 + 3.262e{+}07x^5 - 3.33e{+}07x^4 + 1.529e{+}07x^3 \\ \quad - 3.355e{+}06x^2 + 4.429e{+}05x + 2617 \end{cases} \tag{3.4}$$

### 3.1.3 Introduction of Migration Factors and Online Migration of the Model

The onboard lithium-ion battery applied to new energy vehicles itself is a highly nonlinear complex system, with complex and changeable internal chemical reactions and high instability, and easily affected by ambient temperature and aging (Babaeiyazdi, Rezaei-Zare, and Shokrzadeh 2021; Yinfeng Jiang and Song 2023). Although the existing BMS has the temperature control function of the battery pack (Gao et al. 2021; Kumari, Singh, and Kumar 2023; Yiqun Liu 2022; Shang et al. 2020), which can significantly reduce the impact of the ambient temperature (Y. Zhou, Deng, and Li 2023), it still cannot avoid the intense uncertainty due to battery aging.

The concept of the migration battery model (Tang et al. 2020; Xiong et al. 2022; R. Xu, Wang, and Chen 2022) is different from the conventional aging battery model. In the traditional modeling process, battery aging is usually not considered a dynamic change factor but is always regarded as static (Gun'ko et al. 2020). To reduce the impact of aging, multiple sets of experiments are typically carried out in different aging states of the battery, obtaining data under different aging states and identification of offline or online parameters (Honrao et al. 2021; Xiaoming Liu, Hu, Suo, et al. 2020; Lu et al. 2020; T.-C. Pan et al. 2022). This method will take a lot of time and waste a lot of computing resources. In the process of battery migration modeling (Junhao Qiao et al. 2023; Sheng et al. 2022; Y. Zhang, Liu, et al. 2022; Z. Zhang, Sun, et al. 2023), battery aging is a time-varying dynamic factor to process through the initial model of the offline part, combined with the online migration model, using only the initial condition of battery single condition of experimental data, and the battery in the subsequent work online data can complete the establishment of dynamic migration battery model.

To establish a lithium-ion battery model that can accurately describe different aging states, in this study, the battery equivalent circuit Thevenin model was established as the initial model (Babu 2022; Yongcun Fan et al. 2021; Kong, Wang, and Ping 2021), and the dynamic migration model was built, as shown in Figure 3.3.

In the Figure 3.3 Thevenin model, $U_{OC}$ indicates the open circuit voltage, and the $R_0$ represents the ohmic internal resistance inside the battery, formed by the resistance of the material inside the battery and the contact between the fabric. $R_P$ represents the polarization resistance, $C_P$ represents the polarized capacitance, and $R_P$ and $C_P$ are the parallel circuits that describe the polarization process inside the battery. $U_L$ is the closed circuit voltage formed after the battery is connected to the external circuit. Based on the analysis of the equivalent

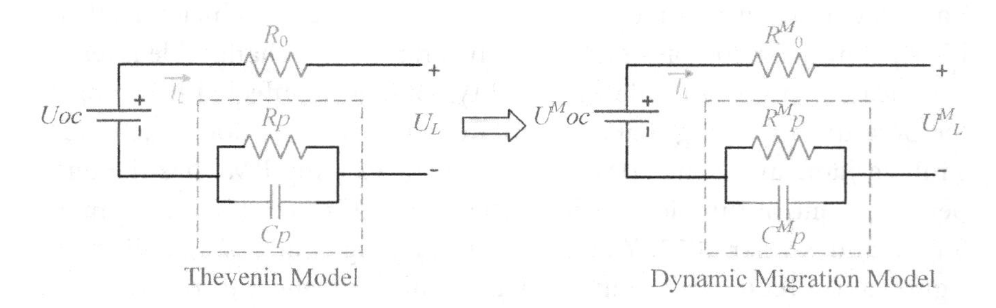


FIGURE 3.3 Dynamic migration cell modeling.

circuit composition, according to Kirchhoff's voltage law (KVL) (Deng, Ye, and Huang 2023; Kahnamouei and Lotfifard 2023; Tong et al. 2023), the state space equation and observation equation of SOC and SOH cooperative estimation based on the Thevenin model are obtained, as shown in Equation (3.5).

$$\begin{cases} \begin{bmatrix} SOC_{k+1} \\ U_{p,k+1} \end{bmatrix} = \begin{bmatrix} 1 & 0 \\ 0 & e^{-\Delta T/\tau} \end{bmatrix} \begin{bmatrix} SOC_k \\ U_{L,k+1} \end{bmatrix} + \begin{bmatrix} -\dfrac{\Delta T}{Q_k} \\ R_p\left(1-e^{-\Delta T/\tau}\right) \end{bmatrix} I_k + \begin{bmatrix} w_{1,k} \\ w_{2,k} \end{bmatrix} \\ Q_{k+1} = Q_k + r_k \\ U_{L,k+1} = U_{OC,k+1} - U_{p,k+1} - IR_0 + v_k \end{cases} \tag{3.5}$$

In Equation (3.5), $[SOC\ U_P]^T$ is set as the system state variable, and $\Delta T$ is the sampling time, which is designated as 0.1 s, that is, the battery end voltage is sampled every 0.1 s. $\tau$ is the time constant, and $\tau = R_P C_P$ $w_{1,k}$ and $w_{2,k}$ represent the process noise in the SOC estimation and the polarized voltage estimation, respectively. $r_k$ represents the process noise in the SOH estimation, and $v_k$ represents the observed noise. $Q_k$ represents the current actual capacity of the battery, usually obtained from the capacity calibration experiment, where $k$ indicates the current time point and $k$ + 1 suggests the next time point. The coefficient matrix of the state–space equations and the observation equations is shown in Equation (3.6).

$$\begin{cases} A_k^{SOC} = \begin{bmatrix} 1 & 0 \\ 0 & e^{-\Delta T/\tau} \end{bmatrix} \\ B_k^{SOC} = \begin{bmatrix} -\dfrac{\Delta t}{Q_k} \\ R_p\left(1-e^{-\Delta T/\tau}\right) \end{bmatrix} \\ A_k^Q = 1 \\ C_k^{SOC} = \begin{bmatrix} \dfrac{\partial U_{OC}}{\partial SOC} & -1 \end{bmatrix} \\ C_k^Q = -\dfrac{i\Delta T}{Q_k^2} \end{cases} \tag{3.6}$$

The internal resistance and OCV value of the battery increase with its continuous aging (S. Lee 2021; Wenhua Li et al. 2019). To realize the model migration of the initial cell model in different aging states, the parameters of the initial cell model and the SOC model should be dynamically adjusted at each sampling point (Wenhua Li et al. 2019; B. Pan et al. 2020; Rumberg et al. 2019). Since the initial model relationship curve of migration is a function of the battery's internal parameters and the battery SOC (Jinlei Sun, Tang, et al. 2022; B. Yuan et al. 2022; Yunhong Che, Deng, et al. 2022; Kailong Liu, Tang, et al. 2022; Junhao Qiao et al. 2023) and the SOC obtained in the estimation process are not accurate, we should not only correct the parameters of the function expression but also correct the initial estimated non-exact SOC value, to synchronously improve the accuracy of the final estimated results. Therefore, in the circuit structure diagram of the migration model shown in Figure 3.3, the superscript *M* is used to represent the parameters of the battery model after the migration. Although the structure of the circuit elements constituting the migration model is not different from that of the Thevenin battery equivalent circuit model, the difference is that the amount to be estimated in the migration model is the online migration factor matrix composed of the parameters contained in the relationship curve between the SOC and the parameters and the SOC value itself (Raijmakers et al. 2020; X. Wang and Tong 2023; Zihrul, Lippke, and Kwade 2023), and the state expression is shown in Equation (3.7).

$$\begin{cases} X = \left[x_1, x_2, x_3, \cdots x_{10}\right] \\ SOC_k^M = x_1 SOC_k + x_2 \\ U_{OC,k}^M = x_3 f_{OCV}\left(SOC_k^M\right) + x_7 \\ R_{0,k}^M = x_4 f_{R_0}\left(SOC_k^M\right) + x_8 \\ R_{P,k}^M = x_5 f_{R_P}\left(SOC_k^M\right) + x_9 \\ C_{P,k}^M = x_6 f_{C_P}\left(SOC_k^M\right) + x_{10} \\ U_{L,k}^M = U_{OC,k}^M - U_{P,k}^M - IR_{0,k}^M + v_k \end{cases} \tag{3.7}$$

In the migration model expression in Equation (3.7), $X = [x_1, x_2, x_3, \ldots x_{10}]$ is the online migration factor matrix of the model; $SOC_k{}^M$ is the corrected SOC value; $UOC^M{}_k$, $R^M{}_{0,k}$, $R^M{}_{P,k}$, and $C^M{}_{P,k}$ are the migrated parameter values of the relationship curve between SOC and the battery model parameters; and $U^M{}_{L,k}$ is the end voltage value obtained based on the observation equation of the migration model.

## 3.2 ONLINE PARAMETER IDENTIFICATION BASED ON THE DEVIATION COMPENSATION STRATEGY

### 3.2.1 Recursive Least Squares Method

The recursive least square (RLS) (C. Ge, Zheng, and Yu 2022; C.-S. Huang 2023; Ke Liu, Wang, Yu, et al. 2022; Q. Ouyang, Chen, and Zheng 2020) method is an iterative recursive algorithm based on adaptive filtering theory. This method can be applied to the system parameters of the system model, and parameters are greatly affected by external conditions and can accurately capture the real-time characteristics of the system (Jialu Qiao et al. 2021; M. Zhang, Wang, et al. 2023; T. Zhu, Wang, Fan, et al. 2023). For the system model to be identified, the discrete equation and the corresponding difference equation are shown in Equations (3.8).

$$\begin{cases} G(z)=\frac{y(z)}{u(z)}=\frac{b_1z^{-1}+b_2z^{-2}+\cdots+b_nz^{-n}}{1+a_1z^{-1}+a_2z^{-2}+\cdots+a_nz^{-n}} \\ y(k)=-\sum_{i=1}^{n}a_iy(k-i)+\sum_{i=1}^{n}b_iu(k-i)+v(k) \end{cases} \tag{3.8}$$

Where $a$ and $b$ are the parameters to be estimated, $y(k)$ is the observed value at time $k$ output by the system, $U(k)$ is the $k$ time value input by the system, and $V(k)$ is the random noise with a mean value of 0.

Let $h(k) = [-y(k-1)\ldots, y(k-n), u(k-1), \ldots, u(k-n)]$, and $\theta = [a_1, a_2\ldots, a_n, b_1, b_2\ldots, b_n]^T$, $V_m = [v(1)\ v(2)\ldots v(3)]^T$, where $\theta$ is the parameter to identify. The matrix form of the difference equation in Equation (3.8) is shown in Equation (3.9).

$$Z_m = H_m\theta + V_m \tag{3.9}$$

For the preceding equation, the idea of recurrent least squares is to find an estimate of $\theta$ so that the measurement value $Z\hat{\theta}_I\hat{Z}_i = H_i\hat{\theta}$ and the smallest sum of the difference between estimates is obtained by estimation, as shown in Equation (3.10).

$$J\left(\hat{\theta}\right)=\left(\hat{Z}_m-H_m\hat{\theta}\right)^T\left(Z_m-H_m\hat{\theta}\right) \tag{3.10}$$

According to the extreme value theorem, finding the minimum value of the preceding equation is equivalent to finding its derivative and then solving it. That is, the least squares estimate of $\theta$ is shown in Equation (3.11).

$$\hat{\theta}=\left[H_m^TH_m\right]^{-1}H_m^TZ_m \tag{3.11}$$

The idea of the RLS algorithm is to use new observations to correct the estimates obtained based on the latest estimation results until satisfactory accuracy. It considers that the measurement data may be obtained under different conditions, the measurement accuracy is affected by many factors, and therefore, the received data may have credibility problems. Thus, each measurement is treated using a weighted method, and the recurrence equation of the least squares method is shown in Equation (3.12).

$$\begin{cases} \hat{\theta}_{m+1} = \hat{\theta}_m + K_{m+1}\left[z(m+1) - h(m+1)\hat{\theta}_m\right] \\ P_{m+1} = P_m - P_m h^T(m+1)\left[W^{-1}(m+1) + h(m+1)P_m h^T(m+1)\right]^{-1} \\ \qquad h(m+1)P_m \\ K_{m+1} = P_m h^T(m+1)\left[W^{-1}(m+1) + h(m+1)P_m h^T(m+1)\right]^{-1} \end{cases} \tag{3.12}$$

Among it, $W_m$ is the weight matrix, which is a symmetric positive definite matrix, usually a diagonal matrix, $P_m = [H_m{}^T W_m H_m]^{-1}$, gain matrix $K_{m+1} = P_{m+1} h^T(m+1) W(m+1)$. The mathematical expression for the Thevenin circuit model was discretized using the double-linear transformation method, as shown in Equation (3.13).

$$\begin{aligned} U_{L,k} - U_{OC,k} &= \left[U_{L,k-1} - U_{OC,k-1} I(k) I(k-1)\right]\left[a \quad b \quad c\right]^T + e(k) \\ &= h^T(k)\theta(k) + e(k) \end{aligned} \tag{3.13}$$

Where, $\theta(k) = [a, b, c]^T$ for the parameter vector to be identified. When the online parameter identification is completed, the actual battery parameters can be obtained as shown in Equation (3.14).

$$\begin{cases} R_0 = \dfrac{c-b}{1+a} \\ R_p = \dfrac{2(c+ab)}{a^2-1} \\ C_p = \dfrac{-T(1+a)^2}{4(c+ab)} \end{cases} \tag{3.14}$$

### 3.2.2 Deviation Compensation-Recursive Least Squares Method

Because there will inevitably be some noise signals in the battery measurement data, the traditional RLS algorithm cannot identify the battery model parameters with high accuracy, and the accuracy of the battery model parameters directly affects the co-estimation accuracy of the battery SOC and SOH (Jialu Qiao et al. 2021; M. Zhang, Wang, et al. 2023; T. Zhu, Wang, Fan, et al. 2023). Based on the inherent defects of RLS, this research proposes the bias compensation recursive least square (BCRLS) algorithm based on bias compensation. It introduces noise variance estimation, which can compensate for the parameters identified by traditional RLS and realize the accurate identification of battery model parameters (Ke Liu, Wang, Yu, et al. 2022; X. Yang, Wang, et al. 2023). The parameter initialization of the BCRLS is as described in Equation (3.15).

$$\begin{cases} \hat{\theta}_c(0) = \theta(0) = \varepsilon \\ J(0) = 0 \\ P(0) = \delta I_0 \end{cases} \tag{3.15}$$

$\hat{\theta}_c(0)$ in the preceding equation is the initial value of the deviation compensation parameter; $\theta(0)$ is the initial value of the RLS for parameter identification, the initial value of the $J(0)$ error covariance function; and $P(0)$ is the initial value of the covariance matrix. $\delta$ is usually a large positive number, and $I_0$ is an identity matrix. The prediction output and estimation error of the model are shown in Equation (3.16).

$$\begin{cases} \hat{y}(k) = \phi^T(k)\theta(k-1) \\ e(k) = y(k) - \hat{y}(k) \end{cases} \tag{3.16}$$

Among it, $\varphi^T(k) = [-y(k-1), -y(k-2) \ldots -y(k-n_a), u(k-1), \ldots, u(k-n_b)]$, $\theta(k)$ is the parameter vector to be identified. In each iteration, the algorithm uses the difference between the systematic observed calculated values and the actual observations as well as the gain $K$ to correct the final estimate. The calculation of the gain matrix $K$ and the parameter estimation of the battery is shown in Equation (3.17).

$$\begin{cases} K(k) = P(k-1)\phi(k)\left[1 + \phi^T(k)P(k-1)\phi(k)\right]^{-1} \\ \theta(k) = \theta(k-1) + K(k)e(k) \end{cases} \tag{3.17}$$

The error standard function $J(k)$ is calculated. The estimation of $\sigma^2(k)$ is shown in Equation (3.18).

$$\begin{cases} J(k) = J(k-1) + e^2(k)\left[1 + \phi^T(k)P(k-1)\phi(k)\right]^{-1} \\ \sigma^2(k) = \dfrac{J(k)}{k\left[1 + \theta_c(k-1)\theta(k-1)\right]} \end{cases} \tag{3.18}$$

The covariance matrix $P(k)$ and the parameter $\theta$ of the cell model after deviation compensation of the update of $\theta_c(k)$ are shown in Equations (3.19).

$$\begin{cases} P(k) = \left[I - K(k)\phi^T(k)\right]P(k-1) \\ \theta_c(k) = \theta(k) + k\sigma^2(k)P(k)\theta_c(k-1) \end{cases} \tag{3.19}$$

In this part, the BCRLS can solve, to a certain extent, colored noise system input information interference to the system, which leads to the traditional RLS algorithm of the identification result deviation problem, effectively improving the traditional RLS algorithm of lithium-ion battery model parameter identification accuracy, and realizing the characterization of the battery internal dynamic running state more accurately.

## 3.3 MODEL VALIDATION OF DYNAMIC MIGRATION

### 3.3.1 Comparison and Verification of the Online Model and the Offline Model

To verify the improvement effect of the dynamic migration model proposed in this study on the modeling accuracy compared with the traditional equivalent circuit model, the model accuracy is demonstrated through the HPPC working condition data. Current $I$ is used as the input value of the model to obtain the end voltage of the model output and the actual measured end voltage data of the lithium-ion battery under the same input current. The experimental validation results are shown in Figure 3.4.

As shown in Figure 3.4(b), the output voltage error of the migration model is significantly smaller than the output voltage error of the Thevenin model during the whole charge-and-discharge process. The maximum output error of the Thevenin model is up to −0.0675 V, and that of the migration model is only 0.0451 V. This proves that the migration model can more accurately characterize the dynamics of the battery, with a significant improvement over conventional models. The application of the migration model lays a solid foundation for the better co-estimation of SOC and SOH for lithium-ion batteries.

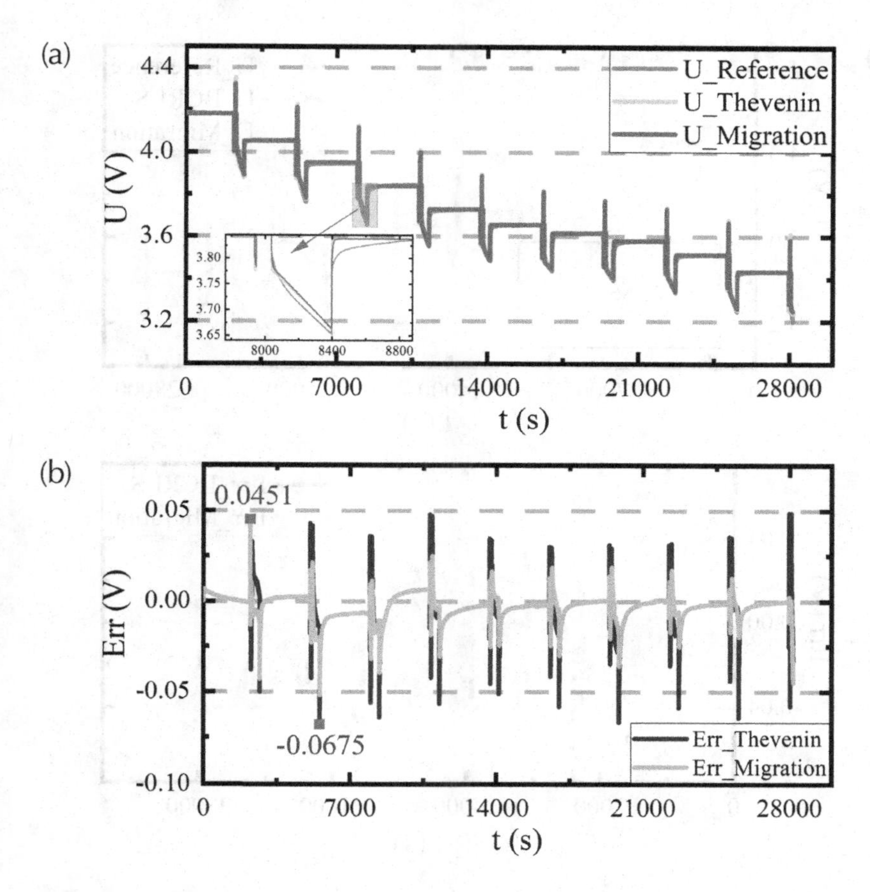


FIGURE 3.4 Comparison of output voltage under HPPC condition: (a) comparison of terminal voltage under HPPC condition; (b) voltage error under HPPC condition.

### 3.3.2 Comparison and Verification of Online Identification and Offline Identification

In this study, due to the construction of the battery migration model, the driver of the migration model needs the structure of the functional relationship between parameters and SOC, so it depends on the offline parameter identification rather than the specific parameter value corresponding to each sampling point obtained by the online parameter identification. However, in general, there are often large deviations in the process of extracting points, curve fitting, and fitting polynomial order selection, and the accuracy of offline parameter identification is often lower than that of online parameter identification. In this study, the functional expression

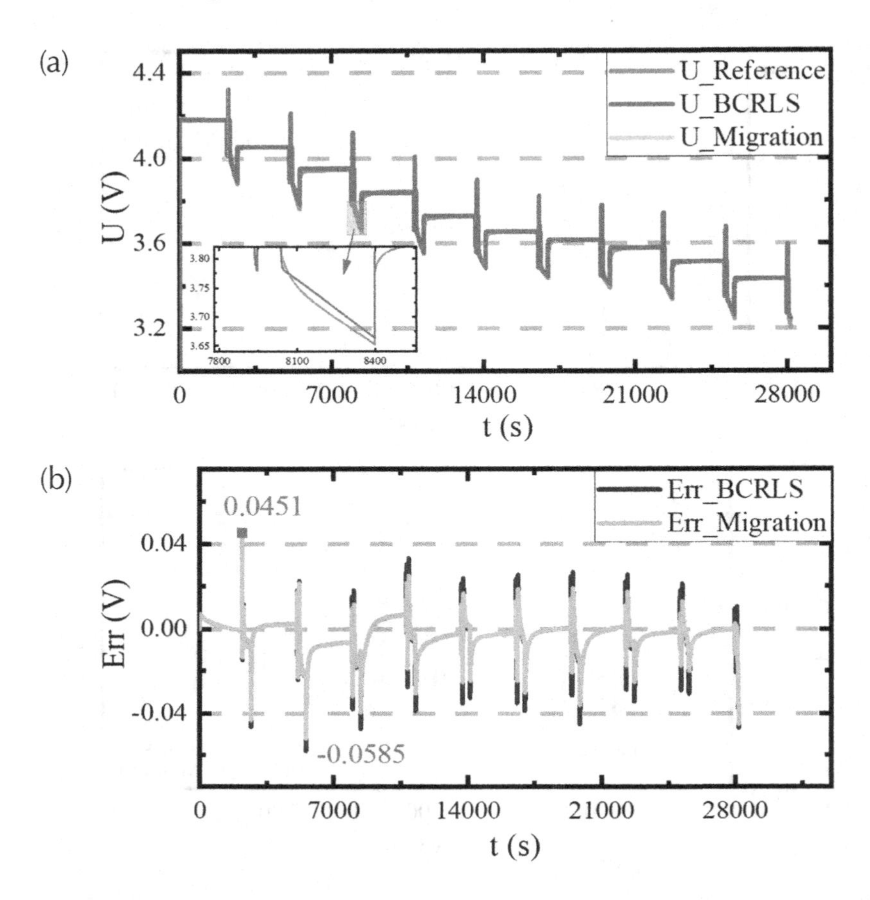


FIGURE 3.5 Comparison of output voltage under HPPC condition: (a) comparison of terminal voltage under HPPC condition; (b) voltage error under HPPC condition.

between SOC and parameters obtained from offline parameter identification is not fixed. Still, it is adaptively adjusted under the particle filter algorithm and the migration factor update. Therefore, the experimental validation results of the online parameter identification and the validity of the parameter identification method in this study are shown in Figure 3.5.

As shown in Figure 3.5(b), during the whole charging and discharge processes, the voltage error of the migration model under the particle filter is less than that of the BCRLS online parameter identification result for the model. The maximum output error of the BCRLS online identification algorithm is −0.0585V, while the migration model is 0.0451V. Compared with offline parameter identification based on the Thevenin model, the advantages of offline parameter identification based on the migration model could be more obvious. However, the accuracy of the model characterization is still better than that of the improved online parameter identification algorithm. This shows that the migration model can more accurately characterize the dynamics of the battery, not only with a significant improvement over the traditional model, but also with higher accuracy over the online algorithms. The application of the migration model lays a solid foundation for improving the accuracy of SOC and SOH in lithium-ion batteries.

## 3.4 SUMMARY OF THIS CHAPTER

In this chapter, dynamic cell migration modeling is introduced by considering the critical factor of cell aging, which affects the regular use of batteries in actual operation. First, the initial Thevenin model is established to obtain the functional relationship between each parameter and the charge state of the battery by offline parameter identification. Then, driven by the particle filtering algorithm, each battery parameter is constantly corrected and updated to realize its dynamic adjustment. By combining the offline initial model part with the online migration model part, the establishment of the active migration battery model is completed by using the small experimental data in the initial state of the battery and the online data in the actual use of the battery. This chapter also introduces the online parameter identification algorithm of the recursive least squares method. It presents the deviation compensation strategy to reduce the influence of colored noise in the traditional recursive least squares method and improve the accuracy of online parameter identification. To verify the characterization of battery dynamic characteristics of the migration model, this chapter will be based on the Thevenin model offline parameter identification and the model validation results of the migration model, and then based on the deviation compensation, the model parameter identification and offline parameter identification of the migration model, which proved the feasibility and effectiveness of the characterization accuracy of the migration model. The precise battery modeling in this chapter lays a solid foundation for the collaborative estimation of SOC and SOH in lithium-ion batteries in follow-up studies.

CHAPTER 4

# SOC and SOH Co-Estimation Based on the Firefly Optimization Algorithm

MOST OF THE EXISTING algorithms are separate estimates for battery SOC or SOH. A double-layer filter is constructed in this study to improve the accuracy of parameter estimation and increase the correlation between parameters. The first-layer filter realizes the SOC estimation based on the particle filter, and the second-layer filter realizes the SOH estimation based on the Kalman filter. Through the mutual correction and secondary feedback of the two, the accuracy of collaborative assessment is improved. Conventional particle filtering in the process of calculation is time-consuming, resampling after particle degradation. This study for the defect introduces population intelligent optimization of the firefly algorithm to improve the particle filtering resampling process and introduces chaos mapping algorithm to enhance the firefly algorithm into local optimal defects, better realize the SOC particles and SOH particle optimization process, and effectively improve its collaborative estimation accuracy.


DOI: 10.1201/9781003486251-4

## 4.1 THE CHAOTIC FIREFLY-PARTICLE FILTERING ALGORITHM

In this study, a particle filter algorithm was used to estimate the SOC of lithium-ion batteries. Aiming at the defects of particle degradation and reduced particle diversity in the traditional particle filtering algorithm, the firefly algorithm of population intelligence optimization is introduced. To effectively improve the dependence of the firefly algorithm on the initial solution, to make the slow convergence rate and the local optimization defects easy to fall into, and to further enhance the accuracy of SOC and SOH collaborative estimation, the chaotic mapping algorithm is introduced to form the chaotic firefly-particle filtering algorithm.

### 4.1.1 The SOC Estimation Model Based on Particle Filtering

PF is a statistical filtering method that transforms the integral operation of Bayesian estimation into a sum operation through Monte Carlo processing to obtain the minimum mean-variance estimate of the system state (Askari et al. 2022; Dong 2020; Kang et al. 2022; Xiao Li, Cheng, and Lin 2021). The basic idea is to constantly adjust the weights and positions of the particles by collecting random samples to correct the previous empirical condition distribution. Compared with other filtering methods such as KF, EKF, and UKF, PF does not have to make any a priori assumptions on the system state and theoretically can be applied to any random system that state-space models can describe. At the same time, KF is only suitable for linear noise, and the modified EKF is only ideal for Gaussian noise (Michalski et al. 2021; Porter 2021; Vitetta et al. 2020). Second, EKF improves the Kalman filter algorithm in the process of solving the noise problem, the state noise and the mean noise, and the variance of observation data. The algorithm of the programming process is usually artificially set and adjusted (W. Xia, Sun, and Zhou 2020; Yeong et al. 2020; Z. Zhou et al. 2021) because these data in the experimental environment and practical applications are challenging to measure, and PF only observes the variance of noise that can be estimated.

In the iteration process of the particle filtering algorithm, a large number of particles are required to participate in the SOC calculation (Huanhuan Li, Qu, et al. 2022). Due to the high randomness in the particle initiation process, some SOC particles and the absolute SOC values will have a large deviation. Then, the phenomenon of particle weight imbalance will appear (Y. Geng, Pang, and Liu 2022; Hao et al. 2023; K.-J. Lee, Lee, and Kim 2023). That is, some particles have a higher weight, some particles have a lower weight, and some seriously deviate from the actual value. Such an unbalanced state will seriously affect the estimation accuracy of the algorithm (Xingtao Liu, Zheng, et al. 2020; Lou 2021; Pang et al. 2022; Qiu, Wu, and Wang 2020; X. Shen et al. 2023). There are also some particles with a minimal weight. In the process of calculation, many small-weight particles that do not play a prominent role in the whole also participate in the analysis, which wastes a lot of time and also increases the amount of computation.

In general, the effectiveness of particles and the degree of particle degradation are inversely correlated, which is often weak in a system with a high adequate particle number. According to different application scenarios and algorithm requirements, an appropriate value can be artificially set as the effective particle threshold. Assuming that the actual number of effective particles is lower than the effective particle threshold in the operation of the algorithm, it is regarded that it cannot meet the requirements, and some methods need to be used to suppress particle degradation (Jingrong Wang, Meng, et al. 2023; S. Wang, Jia, et al. 2023; J. Wu, Xu, and Zhu 2023). Resampling is a common and effective way to suppress particle degradation, which is to resample the particle approximation of the posterior probability density of the system. In the process of resampling, the particles with high weights are copied, while the particles with low weights are not selected. The particles with high weights are simulated to generate a new particle set to reduce the purpose of particle degradation.

PF algorithm is more excellent than other methods in predicting and tracking dynamic parameters and can realize accurate tracking of dynamic parameters (T. Wu, Liu, et al. 2022; C. Yu et al. 2022). PF does not require the chance of random events, which is also a prominent advantage of this algorithm. At present, the algorithm has been widely used in many fields. The PF algorithm used for the iterative calculation process of SOC estimation in lithium-ion batteries is shown in Table 4.1.

In the operation process of PF, the frequency of resampling is usually increased to improve the accuracy of estimation. Suppose the number of resamplings is frequent to a certain extent. In that case, it will cause a large number of computing resources and the waste of computing time and also cause particle dilution, that is, the reduction of particle diversity.

TABLE 4.1 Lithium-Ion Battery SOC Estimation Process Based on the PF Algorithm

1. Initialization: using the prior probability P $(x_0)$ N SOC initial particles.
   Step 1: Initial particle: $\{SOC_0^i\}^N_{i=1}$
   Step 2: Particle weight: $\{q_0^i\}^N_{i=1} = 1/N$
2. Particle state update: according to the system update equation, the prior probability sample of the next moment is obtained, and the particle weight is updated simultaneously.
   Step 1: Prior probability sample at the next time: $\{SOC_k^i\}^N_{i=1}$
   Step 2: Particle weight: $\omega_k^i = \omega_{k-1}^i p\,(U_{L\,(k)}|SOC_k^i)$ i = 1, 2, . . . N
3. Weight normalization: after the system gets the new observed value, through the equation of state to produce a new particle set, and then through the observation equation of the observed value, calculate the observed, and each particle predicted error between the weight of the particles by error will update the state of N SOC particles after normalized weight.
   Step 1: Particle weight: $\omega_k^i$
   Step 2: Normalized weight: $\omega_k^i = \omega_k^i / \Sigma^N_{i=1} \omega_k^i$
4. Resampling: the new random sample distribution generated in the previous step is used to calculate the effective number of particles and determine whether the particles meet the resampling conditions. If the effective number of particles is less than the set effective number threshold, resampling is conducted to obtain a new set of particles.
   Step 1: Effective number of particles: $N_{eff} = 1/\Sigma^N_{i=1}(\omega_k^i)^2$
   Step 2: Judgment condition: $N_{eff} <= N_s$
   Step 3: The weight of each particle in the new particle set after resampling: 1/N
5. Predict the particle state at the next moment: output an estimate based on the particle set and particle weights obtained after resampling.
   Estimated value: $SOC_{k+1}^i$
6. Determine whether the iteration is over; if not, execute k = k + 1, and repeat step (2)–(5).

## 4.1.2 The Firefly Algorithm Optimizes the Resampling Process

To address the problem of reduced particle diversity due to the resampling process in the PF algorithm, a firefly optimization algorithm (Behera et al. 2022; Z. Cheng, Song, et al. 2023; Dif et al. 2020; Mashhour et al. 2020) was introduced. The firefly algorithm simulates the nature of real fireflies between brightness attraction behavior to near the brightest individual population intelligent optimization algorithm (K. Rezaei and Rezaei 2022; S. Tan, Zhao, and Wu 2023; Chunyu Xu, Meng, and Wang 2020; G. Xu, Zhang, et al. 2020). Its principle, when applied to SOC and SOH particles for an optimal process, realizes the particle to the optimal value (the closest to the actual state value), namely, the state of the real value.

In nature, individual fireflies communicate with surrounding individuals by their fluorescent behavior, but the firefly fluorescence is only visible within a certain range (S. Yu, Zuo, et al. 2021). The basic idea of the firefly algorithm is that by randomly initializing the firefly population in a given search space, different individual positions, and different fluorescence brightness, high-brightness fireflies attract low-brightness fireflies to move toward it, during which updates of individual positions occur (Zitouni, Harous, and Maamri 2021). With multiple moves, almost all fireflies are concentrated around the brightest firefly individuals to achieve the optimization process. Therefore, brightness, attraction, and individual position are the three main elements of the algorithm. The relative fluorescence brightness of firefly $i$ versus firefly $j$ is shown in Equation (4.1).

$$I_{ij} = I_0 \times e^{-\gamma \times r_{ij}^2} \tag{4.1}$$

In Equation (4.1), $I_0$ is the maximum fluorescence brightness of firefly $i$. The better the value of the objective function, the higher the brightness of the firefly itself. Applied to the PF algorithm to realize the SOC estimation of the lithium-ion battery, that is, the higher the particle weight, the higher the particle brightness. $\gamma$ is the light intensity absorption coefficient, which was set to 0.98 in this study. $r_{ij}$ is the spatial distance between firefly $i$ and $j$. When this parameter is applied to the PF algorithm, it is considered to be the difference between the estimated values of the SOC particles $i$ and $j$. The relative attraction of firefly $i$ to $j$ is shown in Equation (4.2).

$$\beta_{ij} = \beta_0 \times e^{-\gamma \times r_{ij}^2} \tag{4.2}$$

In Equation (4.2), $\beta_0$ is the attraction degree at the light source ($r = 0$), which is the maximum attraction degree of the light source firefly. The position updates of attracted firefly $j$ and globally optimal firefly $i$ are shown in Equation (4.3).

$$\begin{cases} x_j = x_j + \beta_{ij} \times (x_j - x_i) + \alpha \times (rand - 0.5) \\ x_i = x_i + \alpha \times (rand - 0.5) \end{cases} \tag{4.3}$$

In Equation (4.3), $x_i$ and $x_j$ represent the spatial position of fireflies $i$ and $j$. When applied in the PF algorithm to implement the SOC estimation of lithium-ion batteries, they represent the SOC values of particles $i$ and $j$ in the PF algorithm. The step length factor $\alpha$ is a constant on [0,1], and its value was set to 0.05 in this study. The rand is a random factor on the [0,1] that follows a uniform distribution. Since other fireflies fail to attract the brightest firefly individual, we believe that the individual makes random movements in a smaller regional range to update the position of the most brilliant firefly according to the expression in the second part of Equation (4.3).

### 4.1.3 Chaotic Mapping Passes through the Firefly Optimization Process

*Chaotic* is a complex dynamical behavior of nonlinear systems that uses the properties of chaotic motion to improve the optimal performance of the filtering algorithm (Alhadawi et al. 2020; Aydilek et al. 2021; Das and Saha 2021; Dash et al. 2020). The internal mechanism of chaos thought is to linearly map the optimized variables into a series of chaotic variables using chaos mapping and then optimize the search process according to the ergodicity and randomness of chaos (B.A. Hassan 2021). Finally, the optimal solution obtained from this sequence of chaotic variables is linearly reverse-transformed into the original variable space. After the firefly algorithm completes the particle optimization, PF begins to carry out the particle resampling (Janga and Edara 2021; Kaya et al. 2021; X. Ren, Chen, et al. 2022). In this process, the firefly algorithm has the inherent defects of slow optimization speed and quickly falls into the local optimal (Shaban et al. 2022; Zuoxun Wang, Wang, et al. 2021). Introduce the chaotic mapping algorithm and update the firefly position with the optimal solution generated in the chaotic sequence. The algorithm jumps out of the local optimum.

There are many kinds of commonly used one-dimensional chaotic mapping functions, as shown in Table 4.2.

The cubic map is one of the most common and straightforward chaotic maps. The cubic chaotic map value is between [0, 1], and the distribution

TABLE 4.2 One-Dimensional Chaotic Mapping Functions Commonly Used

| Chaos Mapping Function | Function Expression |
|---|---|
| Cubic shine upon | $z_{n+1} = \rho z_n \left(1 - z_n^2\right)$ |
| Logistic mapping | $z_{n+1} = \mu z_n \left(1 - z_n\right),\ z_0 \notin (0, 0.25, 0.5, 0.75, 1.0),\ \mu \in [0, 4]$ |
| Tent shine upon | $x_{n+1} = \begin{cases} 2x_n, & 0 \le x_n < 0.5 \\ 2(1 - x_n), & 0.5 \le x_n \le 1 \end{cases}$ |
| ICMIC shine upon | $z_{k+1} = \sin(\alpha / z_k),\ \alpha \in (0, \infty)$ |
| Chebyshev mapping | $z_{n+1} = \cos\left(\varphi \cos^{-1} z_n\right)$ |

is not uniform, which cannot meet the optimization requirements of the firefly algorithm (Alqahtani et al. 2023; Anand and Arora 2020; Arora, Sharma, and Anand 2020). In the logistic chaos map, the value range of μ is (0, 4) when the chaotic phenomenon is more significant than 3.6, resulting in {*xn*}. The content of the sequences is in between the [0, 1]. Logistic chaos maps an empty window and uneven distribution of the shortcomings. The empty window here is a large blank in the chaotic space, but also a reflection of uneven distribution, which will cause the slow optimization speed and low efficiency of particles. Hence, it is not suitable for the improvement of the firefly algorithm (Atali et al. 2021; Q. Cheng, Wang, et al. 2023; Chou and Dinh-Nhat 2020; Ganguli, Kaur, and Sarkar 2020). ICMIC chaotic map is a one-dimensional infinite folding iterative chaotic map, but it has a high requirement for the iterative initial value. That is, the initial value is zero, or a fixed point can produce a chaos effect, so it cannot be used for the optimization of the firefly algorithm (Gudise, Babu, and Savithri 2023; Gumuscu et al. 2022; M.H. Hassan et al. 2023; Jiang Li, Guo, et al. 2021). Chebyshev chaos map is a one-dimensional chaotic map with good nonlinear dynamics characteristics; the control parameter $\Phi$ [2, 6], the Lyapunov index of the map, is positive, showing that in the range of $\Phi$ [2,6], although the Chebyshev chaotic map, the size of the cluttered space will be restricted, and the mapping method for the selection of the optimization function is limited.

Tent chaotic mapping in mathematical theory refers to a piecewise linear mapping in sections, which is also one of the commonly used ways of chaotic mapping (Zhiqiang Liu, Wang, et al. 2023; Hector Migallon, Belazi, et al. 2020; H. Migallon, Jimeno-Morenilla, et al. 2020; Mohanty and Dash 2023; Pierezan et al. 2021). The image of its function looks like the appearance of a tent (B. Zhao, Chen, et al. 2020; X. Zhou et al. 2022). Tent mapping shows good two-dimensional chaotic characteristics in its parameter range, the size of chaotic space is not restricted, and its distribution function is relatively uniform and has strong correlation characteristics (Junhao Qiao et al. 2023; Taleb et al. 2023; Tawhid and Ibrahim 2022; B. Xie, Li, Li, et al. 2023). Its application in particle optimization algorithms shows fast optimization speed and high optimization efficiency (Yan et al. 2021; Yi et al. 2022; Junqi Yu et al. 2020). So tent mapping was chosen to generate chaotic sequences in this study to optimize the process of finding the optimal solution in the firefly algorithm. The mathematical expression for the tent mapping is shown in Equation (4.4).

$$x_{n+1} = \begin{cases} 2x_n, & 0 \le x_n < 0.5 \\ 2(1 - x_n), & 0.5 \le x_n \le 1 \end{cases} \tag{4.4}$$

First, the initial population with a uniform distribution is randomly generated by the tent chaotic map to ensure the randomness and diversity of individual particles, which is beneficial to improve the convergence rate of the algorithm. In one iteration, in the firefly population $X_i$ in ($x$ = 2, 1, . . . N), all fireflies were sorted from large to small according to the individual fluorescence brightness (particle weight), taking the top 5% of the better individuals. The minimum value of $X_{min}$ in this 5% was obtained, and the maximum value of $X_{max}$ as a chaotic search space. One firefly was randomly selected from the first 5% of individuals and assigned its spatial position (SOC value) $X_A$ as a primary solution. Equation (4.5) is used to apply the $X_A$ mapped to the (0,1) interval. Set the initial value of the chaotic sequence *Tent* (m) *Tent* (1) as shown in Equation (4.5).

$$Tent(1) = \frac{(x_A - X_{min})}{(X_{max} - X_{min})} \tag{4.5}$$

Put Equation (4.5) into the tent chaos map of Equation (4.4), and iteratively generate the chaotic variable sequence *Tent* ($m$) ($m$ = 1, 2, . . . , $ITER_{max}$). $ITER_{max}$ is the maximum number of iterations of chaos search, set to $ITER_{max}$ in this study at 200. The resulting sequence *Tent*($m$) is solved back to the original solution space, and the new solution sequence $x(m)$ in the original solution space is generated by Equation (4.6).

$$x(m) = x_A + \frac{(X_{max} - X_{min})}{2} \times 2(Tent(m) - 1) \tag{4.6}$$

The fluorescence brightness (particle weight) of each firefly in the unique solution sequence $x(m)$ is calculated successively to generate the new optimal solution $X_B$ and compared with $X_A$. The particle weights are compared, retaining the optimal solution in the two, to get the algorithm out of the local optimal defect.

The firefly algorithm of population intelligent optimization is introduced to improve the reduction of particle diversity caused by the particle filter resampling process, and the tent chaos mapping algorithm is introduced to easily fall into the local optimization problem, which forms the chaotic firefly-particle filter (and judge whether) algorithm in this study.

## 4.2 DYNAMIC UPDATE OF MIGRATION FACTORS AND SOC ESTIMATION

Based on constructing the dynamic migration model, the migration factors in the migration matrix are needed to correct the migration model parameters and improve the representation accuracy of the migration model. Based on the particle filtering, the CF-PF algorithm proposed in this study is used to improve the optimization, realize the dynamic update of the relationship expression between lithium-ion battery SOC and each battery parameter, and then realize the correction of SOC estimation results in the particle filter. A particle filter, Kalman filter two-layer filter, is constructed to achieve SOC and SOH collaborative estimation.

### 4.2.1 Transfer Factor Update Based on CF-PF

The migration factor matrix is an important parameter matrix for correcting the initial cell model and SOC during the online migration. The determination of the migration factor is a nonlinear and non-Gaussian process, so a nonlinear non-Gaussian PF algorithm is chosen to determine the offset factors online (Liang et al. 2021; Z. Wei et al. 2020; R. Zhao et al. 2022). The expression of the cell dynamic migration model is shown in combination with Equation (3.7), and the migration matrix is $X = [x_1, x_2, x_3, \ldots x_{10}]$ as the system state variable, the battery end voltage is taken as the system observation value, and the load current $I_k$ during the operation of the battery and an inaccurate $SOC_k$ is taken as an input to the system. The equation of state of the system is established, as shown in Equation (4.7).

$$\begin{cases} \vec{X}_k = \begin{bmatrix} x_{1,k-1} \\ x_{2,k-1} \\ \cdots \\ x_{10,k-1} \end{bmatrix} + \begin{bmatrix} v_1 \\ v_2 \\ \cdots \\ v_{10} \end{bmatrix}, \begin{matrix} v_1 \sim N\left(0,\sigma_1^2\right) \\ v_2 \sim N\left(0,\sigma_2^2\right) \\ \cdots \\ v_{10} \sim N\left(0,\sigma_{10}^2\right) \end{matrix} \\ U_{L,k} = U_{OC,k} - U_{P,k} - IR_0 + \omega, \omega \sim N\left(0,\sigma_\omega^2\right) \end{cases} \tag{4.7}$$

In the state equation of the system shown in Equation (4.7), the first part of the equation is the state transfer equation of the CF-PF algorithm, which is used to describe how the ten migration factors change from the current moment state to the next moment state. The second part of the equation is the observation equation of the system, which estimates the SOC value as the input to the equation. On the end voltage $UL$, the $UL$ values

TABLE 4.3 Parameter Configuration of the Dynamic Migration Model

| $i$ | 1 | 2 | 3 | 4 | 5 | 6 | 7 | 8 | 9 | 10 | ω |
|---|---|---|---|---|---|---|---|---|---|---|---|
| $v_{i,0}$ | 1 | 0 | 1 | 1 | 1 | 1 | 0 | 0 | 0 | 0 | *null* |
| $\sigma_i^2$ | 0.00001 | 0.002 | 0.0001 | 0.0001 | 0.0001 | 0.0001 | 0.001 | 0.001 | 0.001 | 0.001 | 0.001 |

versus the actual measurement of the *UL* values were compared to judge the quality of the estimated SOC values. To simulate practical applications, both the state transition and observation equations give reasonable noise. With ten migration factors within the system, the noise was treated with $\sigma 1 \sim 10$ representation; there is only one observation noise, and $\sigma\omega$, to distinguish it from the preceding ten noises, conducts a presentation. The parameter configuration of the dynamic migration model, namely, the migration matrix $X = [x1, x2, x3, \ldots x10]$ variance of the initial and state values, $\sigma i2$, as shown in Table 4.3.

Based on the PF algorithm, the migration factor iterative update process is realized, and the CF-PF algorithm is proposed in this study to improve and optimize, learning the dynamic update of the relational expression between the SOC of lithium-ion battery and each battery parameter, and then realizing the correction of the SOC estimation results in the particle filter.

### 4.2.2 The SOC Estimation Based on the Migration Model

In this study, the estimation of SOC of lithium-ion batteries mainly depends on the improved optimization algorithm based on the particle filtering algorithm. The essence of SOC is estimated by randomly generating a large number of particles. Each developed SOC particle, according to the state transfer equation, introduced the estimate of the next moment, determined the weight of particle estimates, and then relied on the importance to get the final assessment at the next moment. However, in this basic estimation process, the parameters of the state transfer equation and the observation equation are determined by the internal parameters of the battery model. In contrast, the determination of the internal parameters of the battery model depends on the functional relationship between the offline identification and the SOC, which are fixed and do not change with aging.

Integrating the update of the transfer factor into the PF algorithm to estimate the SOC, the system state transfer equation, and $R_0$ in the observation equation, $R_P$, $C_P$, $U_L$, and $U_{OC}$ being replaced with post-migration values, the initial SOC estimates obtained from PF are considered imprecise and thus also migrating it.

$$\begin{cases} SOC_{k+1}^{,M} = SOC_k - \dfrac{\Delta T}{Q_k} I_k + w_{1,k} \\ U_{p,k+1} = e^{-\frac{\Delta T}{\tau}} U_{p,k} + R_p \left(1 - e^{-\frac{\Delta T}{\tau^M}}\right) I_k + w_{2,k} \\ U_{L,k+1}^M = U_{OC,k+1}^M - U_{p,k+1} - IR_{0^M} + v_k \end{cases} \tag{4.8}$$

$SOC_k{}^M$ is the corrected SOC value; $U_{OC}{}^M{}_{,k}$, $R^M{}_{0,k}$, $R^M{}_{P,k}$, and $C^M{}_{P,k}$ are the migrated parameter values of the relationship curve between SOC and battery model parameters; and $U^M{}_{L,k}$ is the end voltage, obtained according to the observation equation of the migration model $\tau^M = R_P{}^M C_P{}^M$. The parameters after the migration change, and the SOC values are used as the update variables of the PF algorithm to determine the weight of the SOC particle estimate. Then, the final assessment of the next moment is obtained based on the weight.

That is, in the SOC estimation process based on the migration model, the original system state variable $[SOC\ U_P]^T$ expands to the migration matrix $X = [x_1, x_2, x_3, \ldots x_{10}]$. The internal parameters of each battery are adaptively adjusted at each sampling point to adapt to different battery aging conditions and obtain more accurate SOC values in line with the actual situation.

## 4.3 SOC AND SOH CO-ESTIMATION BASED ON PF-EKF

### 4.3.1 Construct the Mutual Correction Mechanism of Double Filters

In this study, we combined the PF algorithm and the EKF algorithm to achieve the synergistic estimation of SOC and SOH in lithium-ion batteries. After obtaining the SOC estimate of the current moment through the CF-PF algorithm, the value is used as input to the current moment SOH estimation filter, the second-layer filter, to correct the a priori estimated capacity value Q obtained from the state transfer equation in Equation (3.5). After receiving the capacity Q at the current moment, it is used as input to the CF-PF algorithm, and the SOC estimate at the next moment is calculated by updating the coefficient matrix *B* of the state transfer equation in Equation (3.6). In separate estimates of SOC, *Q* in matrix *B* is the current actual capacity obtained from a capacity calibration experiment, which does not change with the number of iterations. In the synergy estimation, $Q$ is inversely updated and corrected to achieve the mutual influence and facilitation of the two-state quantities, SOC and SOH. The coefficient matrix of the state–space equation and the observation equation under the two-filter co-estimation condition is shown in Equation (4.8).

$$\begin{cases} \begin{bmatrix} SOC_{k+1} \\ U_{p,k+1} \end{bmatrix} = \begin{bmatrix} 1 & 0 \\ 0 & e^{-\Delta T/\tau} \end{bmatrix} \begin{bmatrix} SOC_k \\ U_{L,k+1} \end{bmatrix} + \begin{bmatrix} -\dfrac{\Delta T}{Q_{k\_update}} \\ R_p\left(1-e^{-\Delta T/\tau}\right) \end{bmatrix} I_k + \begin{bmatrix} w_{1,k} \\ w_{2,k} \end{bmatrix} \\ Q_{k+1} = Q_k + r_k \\ U_{L,k+1} = U_{OC,k+1} - U_{p,k+1} - IR_0 + vk \end{cases} \tag{4.8}$$

In Equation (4.8), $Q_{k\_update}$ represents the current moment estimated capacity value obtained in EKF and is no longer a fixed value. The coefficient matrix of the state–space equations and the observation equations is shown in Equation (4.9).

$$\begin{cases} A_k^{SOC} = \begin{bmatrix} 1 & 0 \\ 0 & e^{-\Delta T/\tau} \end{bmatrix} \\ B_k^{SOC} = \begin{bmatrix} \dfrac{\Delta t}{Q_{k\_update}} \\ R_p\left(1-e^{-\Delta T/\tau}\right) \end{bmatrix} \\ A_k^Q = 1 \\ C_k^{SOC} = \begin{bmatrix} \dfrac{\partial U_{OC}}{\partial SOC} & -1 \end{bmatrix} \\ C_k^Q = -\dfrac{i\Delta T}{Q_k^2} \end{cases} \tag{4.9}$$

The mechanism of mutual correction between SOC and SOH by particle filter–Kalman filtering is shown in Figure 4.1.

### 4.3.2 Build a Complete SOC/SOH Co-Estimation Framework

In this study, the overall SOC/SOH co-estimation framework is divided into two parts: the construction of the offline initial model and online

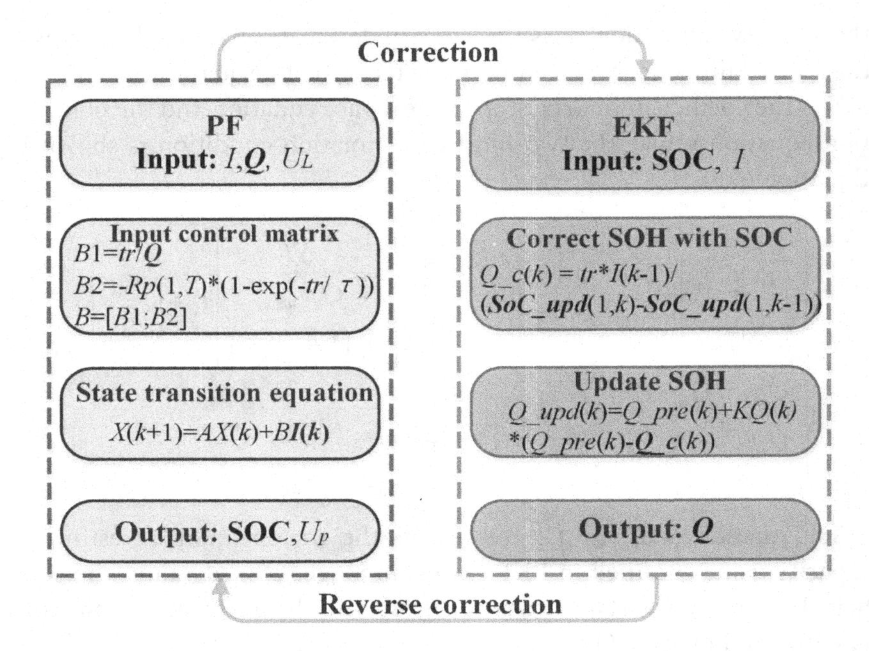


FIGURE 4.1 Mutual correction of the SOC and the SOH.

model migration. In the first part, the initial Thevenin equivalent circuit model is constructed, the offline parameters are identified according to the HPPC condition data, and the functional relationship between SOC and SOC-OCV curve is obtained. That is, the construction of the offline initial model is completed.

In the second part, the firefly algorithm of population intelligence optimization is introduced to improve the optimization accuracy and speed of the traditional PF algorithm and the resampling process. The tent chaotic mapping algorithm is introduced to seek a better optimal solution to help the firefly algorithm jump out of the local optimal. The proposed CF-PF algorithm realizes the iterative update of the migration factor in the battery migration model. It constructs a particle filter–Kalman filter double filter to learn the cooperative estimation and mutual correction of SOC and SOH in lithium-ion batteries. The algorithm's overall process is shown in Figure 4.2.

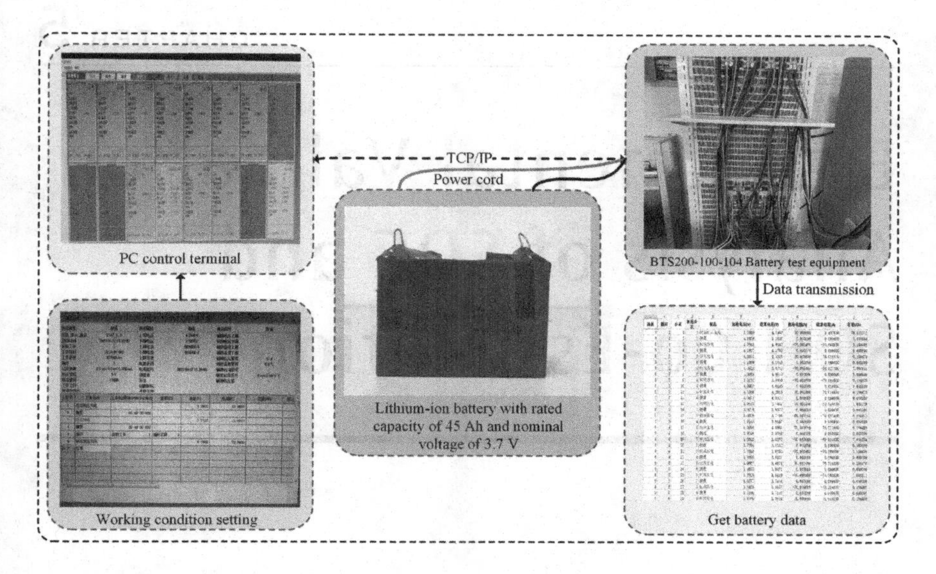


FIGURE 4.2 SOC and SOH are estimated based on the migration model and chaotic firefly-particle filtering algorithm.

## 4.4 SUMMARY OF THIS CHAPTER

In this chapter, a double-layer filter is constructed. The first layer filter realizes SOC estimation based on the particle filter, and the second layer filter realizes SOH estimation based on the Kalman filter. Through the mutual correction and secondary feedback of the two, the purpose of improving the accuracy of assessment estimation is achieved. The conventional particle filter has some problems, such as large calculation amount, long time, and degradation after particle resampling. In this chapter, the firefly algorithm of population intelligence optimization is introduced to improve the resampling process of particle filter, and the chaotic mapping algorithm is introduced to enhance the defect that the firefly algorithm easily falls into local optimum. Through the preceding optimization, the optimization process of SOC particles and SOH particles can be better realized, and the collaborative estimation accuracy can be effectively improved.

CHAPTER 5

# Experimental Validation Analysis of SOC and SOH Co-Estimation

To verify the CF-PF method for the state of charge and health of lithium-ion batteries, this chapter combines the dynamic migration battery model of the CF-PF method under experimental verification. It compares and analyzes the traditional algorithm results to prove the effectiveness of the proposed algorithm. To verify the adaptive regulation effect of the migration model on the aging state of batteries, this chapter also demonstrates the data of battery conditions in different health states with high SOH and low SOH, and the results are compared and analyzed.

## 5.1 OVERALL EXPERIMENTAL DESIGN

This study is based on the application of the battery dynamic migration model to the experimental design and experiments of HPPC, DST, and BBDST at high health state (SOH = 99.11%) (Bobobee et al. 2022; L. Chen, Wang, et al. 2023; Xianyi Jia, Wang, et al. 2022). After obtaining the data, the results of the CF-PF method, traditional PF algorithm, and F-PF method are determined. Since the purpose of building the dynamic cell migration model in this study is to adapt to the dynamic characteristics of batteries in different aging states, it is necessary to verify the batteries in other aging states. After 122 cycle charging and discharge experiments on the ternary lithium-ion battery with a rated


DOI: 10.1201/9781003486251-5

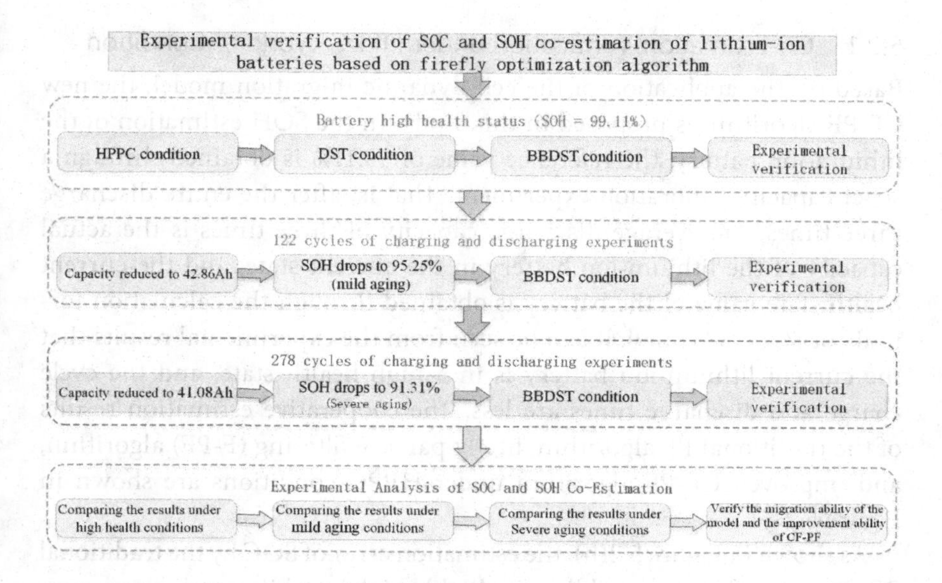


FIGURE 5.1 Overall experimental design framework.

capacity of 45 Ah, the capacity of the battery was reduced to 42.86 Ah, and the current SOH was 95.25%. Then, the BDST condition experiment was conducted. After obtaining the data, the three algorithms were verified and analyzed. After 278 cycle charge and discharge experiments on the ternary lithium-ion battery used in the investigation, the capacity was reduced to 41.08 Ah, and the current SOH was 91.31%. The BBDST experiment was conducted again, and after the data were obtained, the three algorithms were verified and analyzed. Based on this experimental verification design idea, the overall practical design framework is constructed as shown in Figure 5.1.

## 5.2 CO-ESTIMATION VERIFICATION UNDER BATTERY HEALTH STATE

To verify the cooperative estimation accuracy of the CF-PF algorithm for SOC and SOH, the simple HPPC condition and complex BBDST condition were used to test and verify the power lithium-ion battery under the high health state (SOH = 99.11%), and the cooperative estimation accuracy of SOC and SOH under different algorithms was compared through experimental data.

### 5.2.1 Co-Estimation Verification under HPPC Working Condition

Based on the application of the cell dynamic migration model, the new CF-PF algorithm is proposed in this study. In the SOH estimation of the lithium-ion battery, the reference value of 99.11% is obtained through a strict capacity calibration experiment. That is, after the entire discharge three times, the average discharge capacity of three times is the actual capacity of the lithium-ion battery in the current state, and the current health state value of the battery is obtained through the calculation formula of SOH [389, 469]. It can be seen from the experimental results that the current lithium-ion battery is in a high health state, and the cycle charge and discharge times are less. The cooperative estimation results of the traditional PF algorithm, firefly-particle filtering (F-PF) algorithm, and improved CF-PF algorithm under HPPC conditions are shown in Figure 5.2.

As shown in Figure 5.2(b), the estimation error of SOC by the traditional PF algorithm fluctuates wildly, with highly high instability and severe error divergence in the late discharge period, with a maximum error of up to 4.13%. The stability of the F-PF algorithm is significantly improved, and the late error is stable but has slight divergence, with a maximum error of 2.75%, which proves that the introduced firefly algorithm effectively

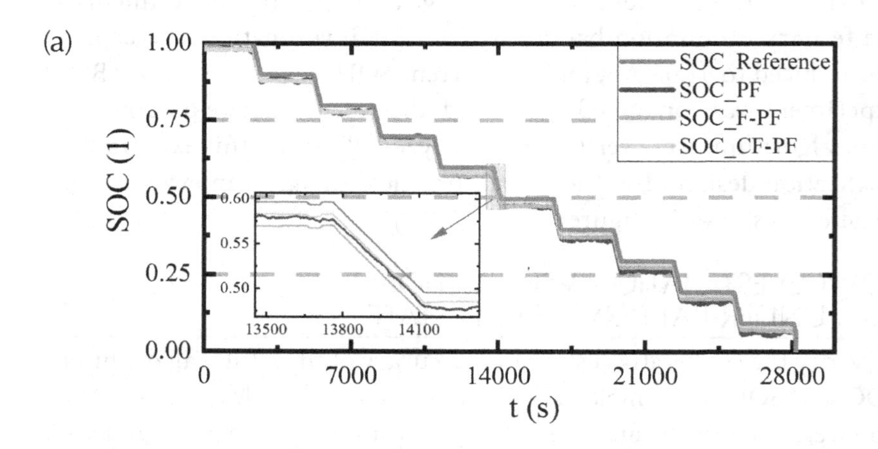


FIGURE 5.2 Comparison of co-estimation results of battery SOC and SOH under HPPC condition: (a) battery SOC estimation results under HPPC operating conditions; (b) battery SOC estimation error under HPPC condition; (c) results of battery SOH estimation under HPPC condition; (d) battery SOH estimation error under HPPC condition.

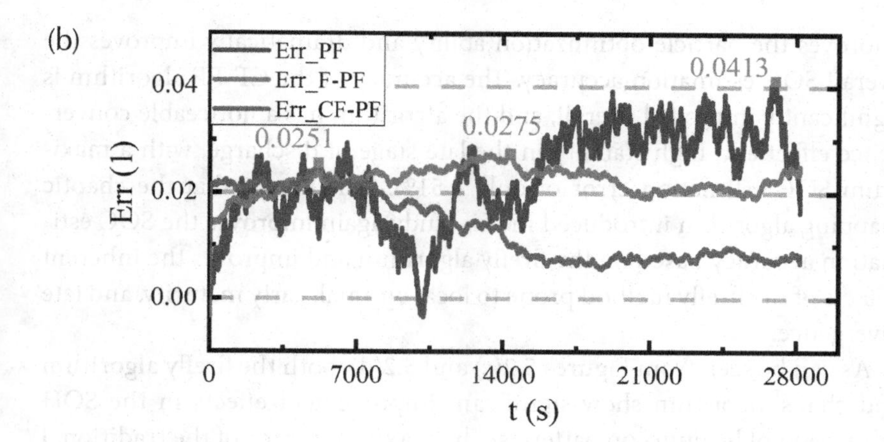


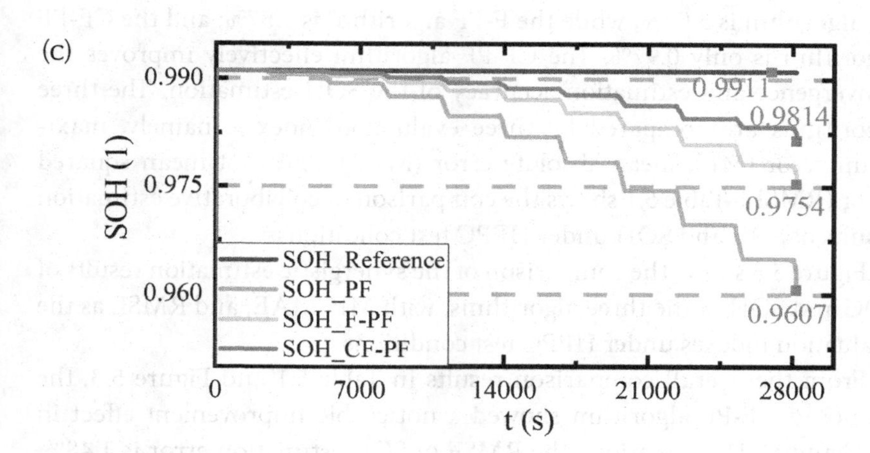


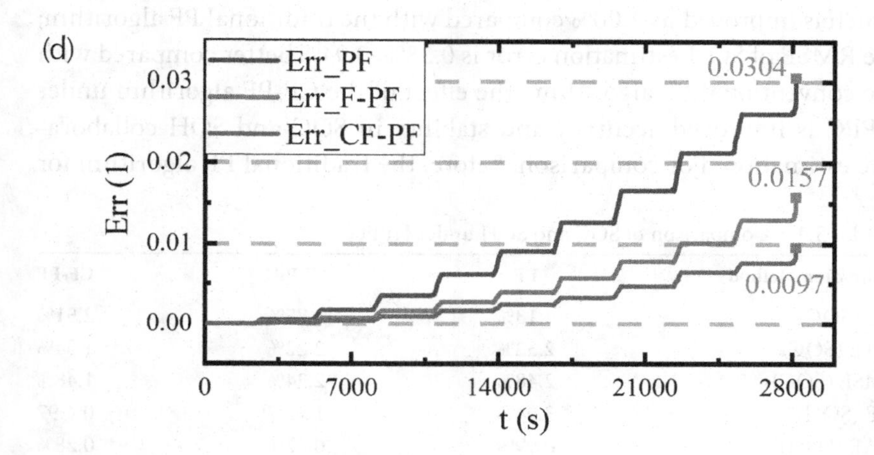


FIGURE 5.2 (Continued)

improves the particle optimization ability and dramatically improves the overall SOC estimation accuracy. The accuracy of the CF-PF algorithm is significantly improved overall, and the algorithm has a noticeable convergence effect and high stability in the late stage of discharge, with a maximum SOC estimation error of only 2.51%. This means that the chaotic mapping algorithm introduced in this study again improves the SOC estimation accuracy based on the firefly algorithm and improves the inherent defects of the firefly method prone to local optimal, early maturity, and late divergence.

As can be seen from Figures 5.2(c) and 5.2(d), both the firefly algorithm and chaos algorithm show significant improvement effects in the SOH estimation of lithium-ion batteries. The maximum error of the traditional PF algorithm is 3.04%, while the F-PF algorithm is 1.57%, and the CF-PF algorithm is only 0.97%. The CF-PF algorithm effectively improves the convergence and estimation accuracy of the SOH estimation. The three algorithms are compared by three evaluation indexes, namely, maximum error (ME), mean absolute error (MAE), and root mean squared error (RMSE). Table 5.1 shows the comparison of collaborative estimation results of SOC and SOH under HPPC test conditions.

Figure 5.3 shows the comparison of the synergistic estimation results of SOC and SOH of the three algorithms, with ME, MAE, and RMSE as the evaluation indexes under HPPC test conditions.

From the overall comparison results in Table 5.1 and Figure 5.3, the proposed CF-PF algorithm showed a noticeable improvement effect in SOC and SOH estimation. The RMSE of SOC estimation error is 1.48%, which is improved by 1.00% compared with the traditional PF algorithm; the RMSE of SOH estimation error is 0.38%, 0.94% better compared with the conventional PF algorithm. The effect of the CF-PF algorithm under HPPC is improved accuracy and stability in SOC and SOH collaborative estimation. For comparison, before, the traditional PF algorithm for

TABLE 5.1 Comparison of SOC and SOH under HPPC

| Estimation Method | PF | F-PF | CF-PF |
|---|---|---|---|
| ME (SOC) | 4.13% | 2.75% | 2.51% |
| MAE (SOC) | 2.32% | 2.22% | 1.34% |
| RMSE (SOC) | 2.48% | 2.24% | 1.48% |
| ME (SOH) | 3.04% | 1%.57 | 0%.97 |
| MAE (SOH) | 0.99% | 0.47% | 0.28% |
| RMSE (SOH) | 1.32% | 0.63% | 0.38% |

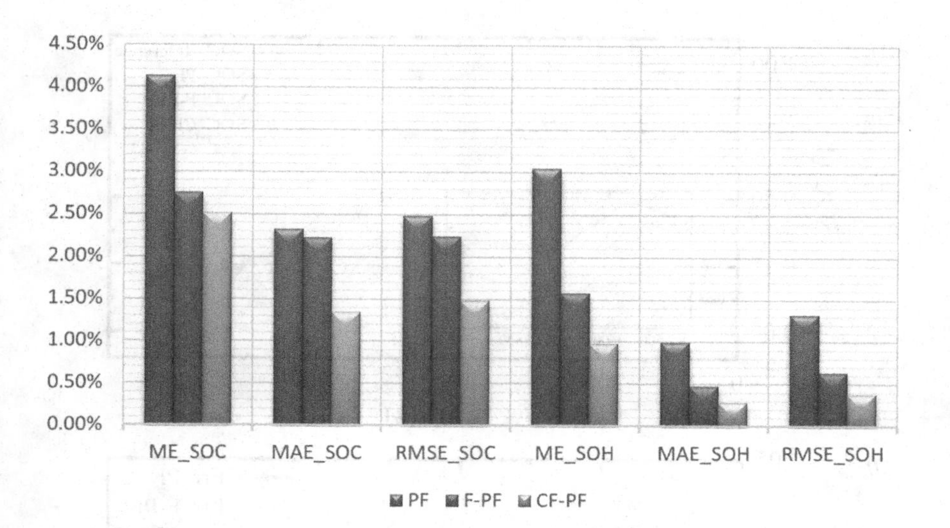


FIGURE 5.3 Comparison of collaborative estimation results of battery SOC and SOH under HPPC condition

SOC and SOH collective estimate running time is 79.8519s; the improved CF-PF algorithm running time is 83.0938s. This is due to joining the intelligent optimization firefly algorithm, the PF algorithm resampling before an additional optimization, and *N* SOC particles compared with the optimal value of "fluorescence brightness" calculation for a move to the optimal value. As a result, compared with the original algorithm, the improved algorithm complexity is slightly improved, but it can significantly improve the accuracy of state estimation. The slight increase in running time still meets the requirements of the algorithm for real-time performance, and the overall enhanced algorithm still shows superior performance.

### 5.2.2 Co-Estimation Verification under DST Working Condition

To further verify the effectiveness of the algorithm proposed in this study for improving the collaborative estimation effect of SOC and SOH, the algorithm is verified under DST working conditions to determine whether the algorithm has the adaptive adjustment ability for more complex working needs. The results of the traditional PF algorithm (Askari et al. 2022; Michalski et al. 2021), the F-PF algorithm without the chaotic-PF algorithm, and the improved CF-PF algorithm under the DST condition are shown in Figure 5.4.

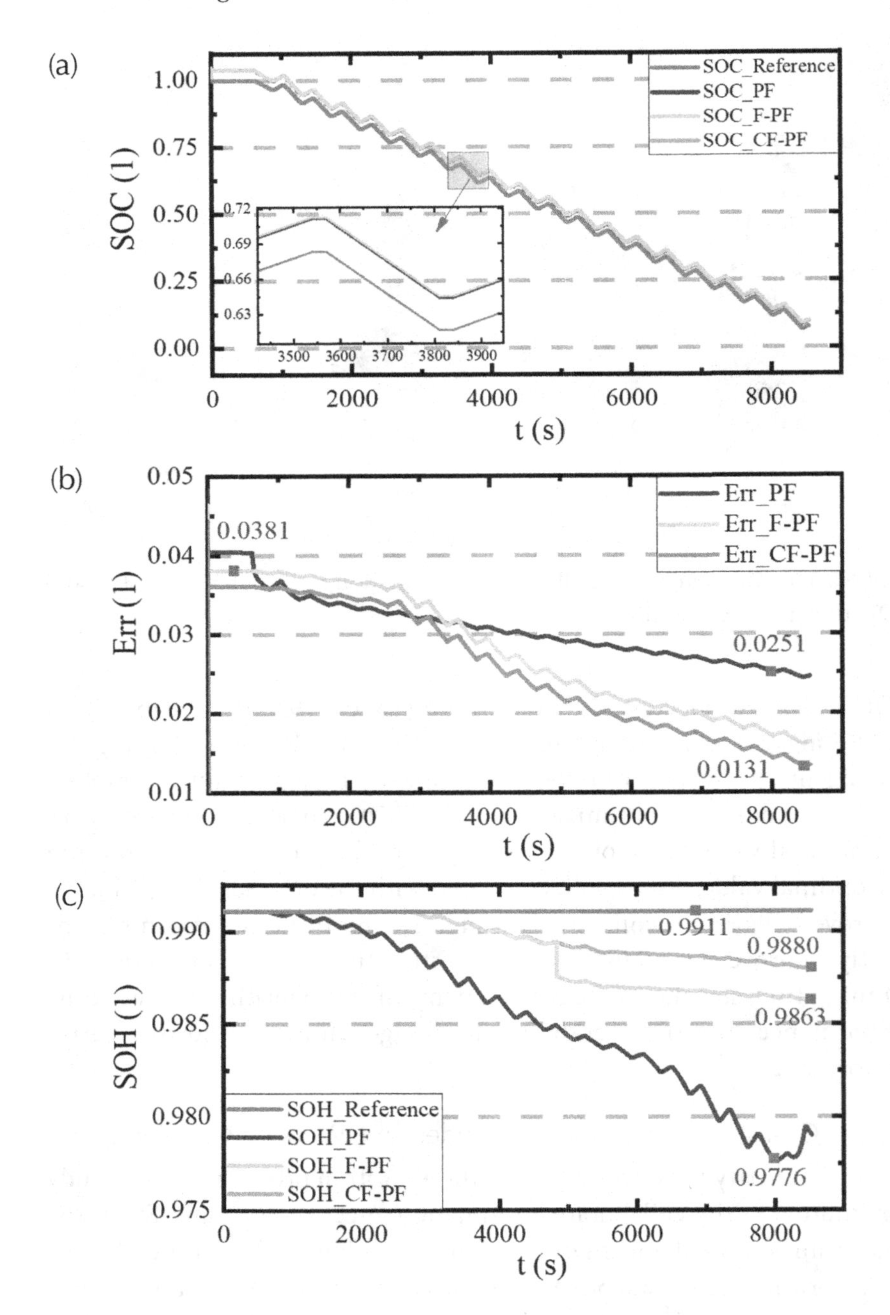


FIGURE 5.4 Comparison of SOC and SOH under DST condition: (a) results of battery SOC estimation under DST condition; (b) battery SOC estimation error under DST condition; (c) SOH estimation results of the battery under DST condition; (d) battery SOH estimation error under DST condition.

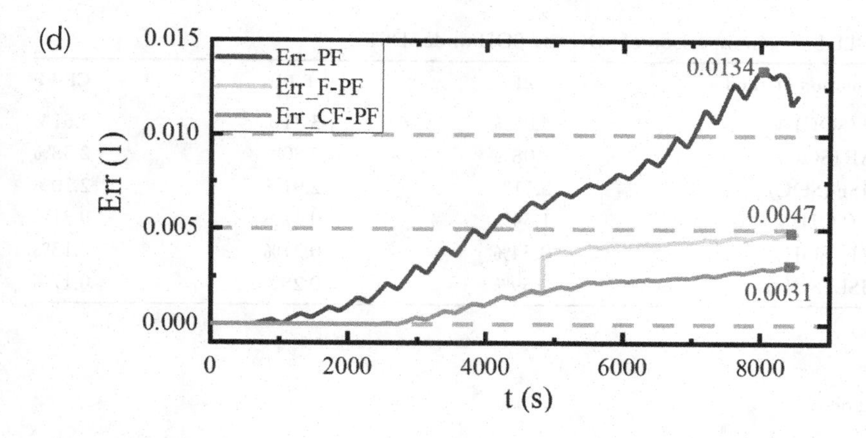


FIGURE 5.4 (Continued)

As shown in Figure 5.4(b), the estimation accuracy of the CF-PF algorithm in the initial stage of discharge is reasonable, and it does not show apparent advantages compared with the other two algorithms. However, in the middle and late stages of release, the estimation error of the F-PF algorithm is better than that of the traditional PF algorithm, and the accuracy of the algorithm is greatly improved. This reflects the practical improvement of the SOC particle optimization process, by introducing the firefly algorithm into the PF algorithm. However, the estimation accuracy of the CF-PF algorithm is significantly improved, and it has high convergence, which can track the actual value well, which reflects the practical improvement of SOC particles by introducing tent chaos mapping into the F-PF algorithm.

As shown in Figure 5.4(d), at the end of the discharge, the estimation error of the traditional PF algorithm on the battery SOH increases sharply, which is related to the solid electrochemical reaction inside the battery itself, indicating that the PF algorithm cannot correct the error increase in the late stage. In this case, the F-PF algorithm and the CF-PF algorithm can still maintain good estimation accuracy, with a maximum F-PF error of 0.47% and a slight increase in the late discharge error. The total error of CF-PF is only 0.31%, and the estimation error in the late discharge remains stable. The superiority of the CF-PF algorithm proposed in this study is verified by comparing the synergistic estimation results of SOC with SOH. The three algorithms are compared by three evaluation indexes: ME, MAE, and RMSE. And Table 5.2 shows the comparison of the synergistic estimation results of SOC and SOH under DST test conditions.

TABLE 5.2 Comparison of SOC and SOH under DST

| Estimation Method | PF | F-PF | CF-PF |
|---|---|---|---|
| ME (SOC) | 4.05% | 3.81% | 3.61% |
| MAE (SOC) | 3.08% | 2.80% | 2.58% |
| RMSE (SOC) | 3.11% | 2.91% | 2.70% |
| ME (SOH) | 1.34% | 0.47% | 0.31% |
| MAE (SOH) | 0.54% | 0.21% | 0.13% |
| RMSE (SOH) | 0.69% | 0.28% | 0.17% |

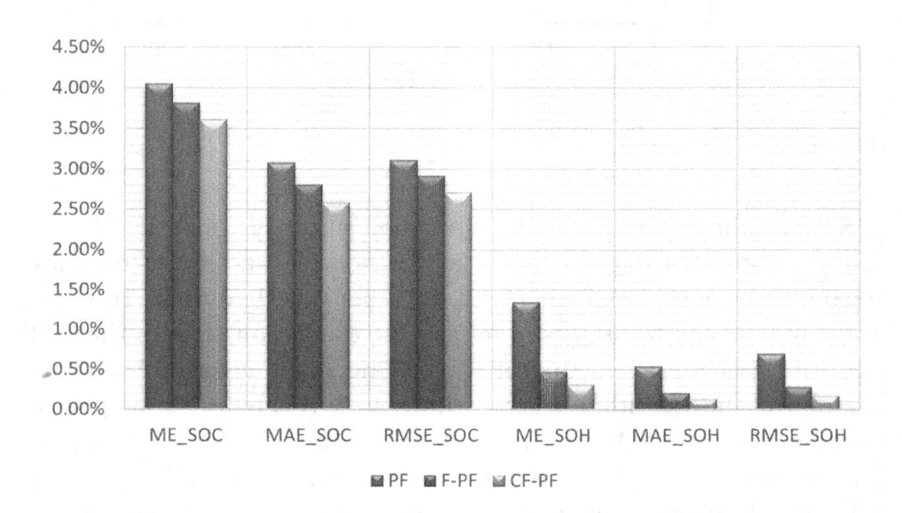


FIGURE 5.5 Comparison of SOC and SOH under DST condition.

Figure 5.5 shows the comparison of the synergistic estimation results of SOC and SOH of the three algorithms, with ME, MAE, and RMSE as the evaluation indexes under DST test conditions.

According to the overall comparison results of Table 5.2 and Figure 5.5, the proposed CF-PF algorithm shows a noticeable improvement effect in the SOC and SOH estimation under DST conditions, and the RMSE of the SOC estimation error is 2.70%, which is 0.41% improvement compared with the traditional PF algorithm; the RMSE of the SOH estimation error is 0.17%, which is 0.52% improvement compared with the conventional PF algorithm. The effect of the CF-PF algorithm in improving the accuracy and stability in SOC and SOH collaborative estimation under DST conditions is verified.

### 5.2.3 Co-Estimation Verification under BBDST Working Condition

The working conditions of new energy electric vehicles in the actual operation are dynamic, complex, changeable, and irregular (Q. Guo, Liu, and

Cai 2023; B. Sun and Ju 2023). To better simulate the natural operation effect of the onboard battery, the ternary lithium-ion battery is tested by setting the charging and discharging steps of BBDST. BBDST is a working condition collected from the actual operation data of Beijing buses, including not only the primary working conditions, such as starting, braking, and parking, but also the operations, such as acceleration, sliding, and rapid acceleration (C. Zhu, Wang, Yu, et al. 2023; T. Zhu, Wang, Fan, et al. 2023). The results of SOC and SOH of the traditional PF algorithm, F-PF algorithm, and improved CF-PF algorithm under BBDST condition are shown in Figure 5.6.

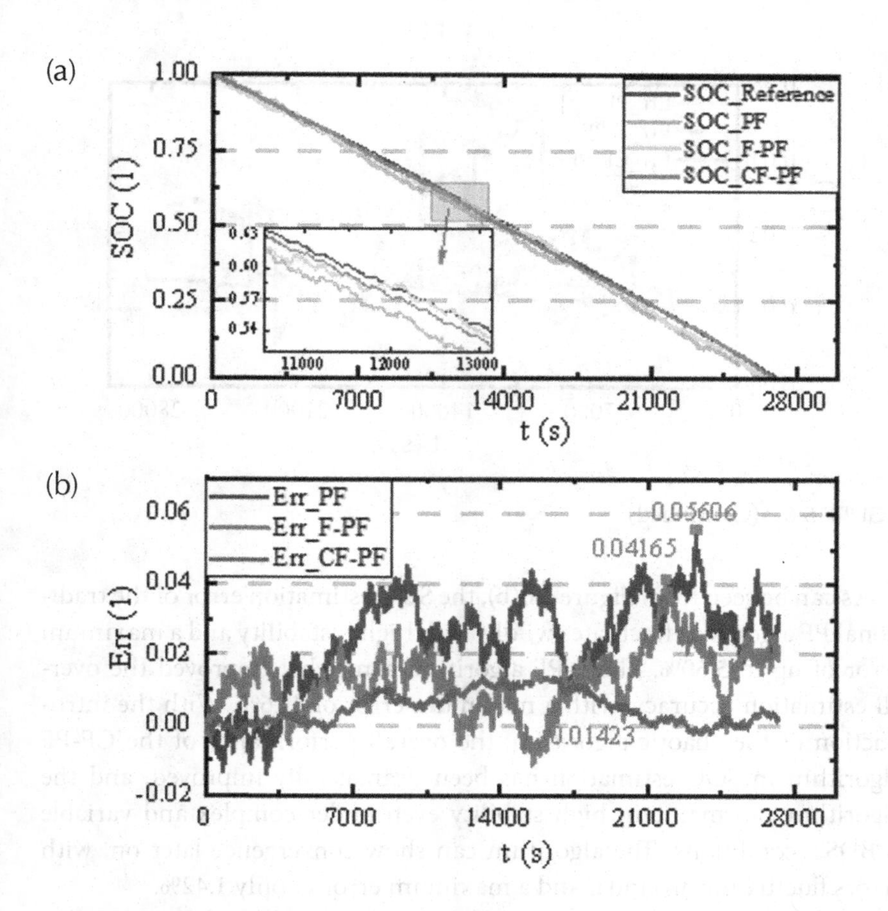


FIGURE 5.6 Comparison of collaborative estimation results of SOC and SOH under BBDST: (a) battery SOC estimation results under BBDST conditions; (b) battery SOC estimation error under BBDST condition; (c) results of battery SOH estimation under BBDST condition; (d) battery SOH estimation error under BBDST condition.

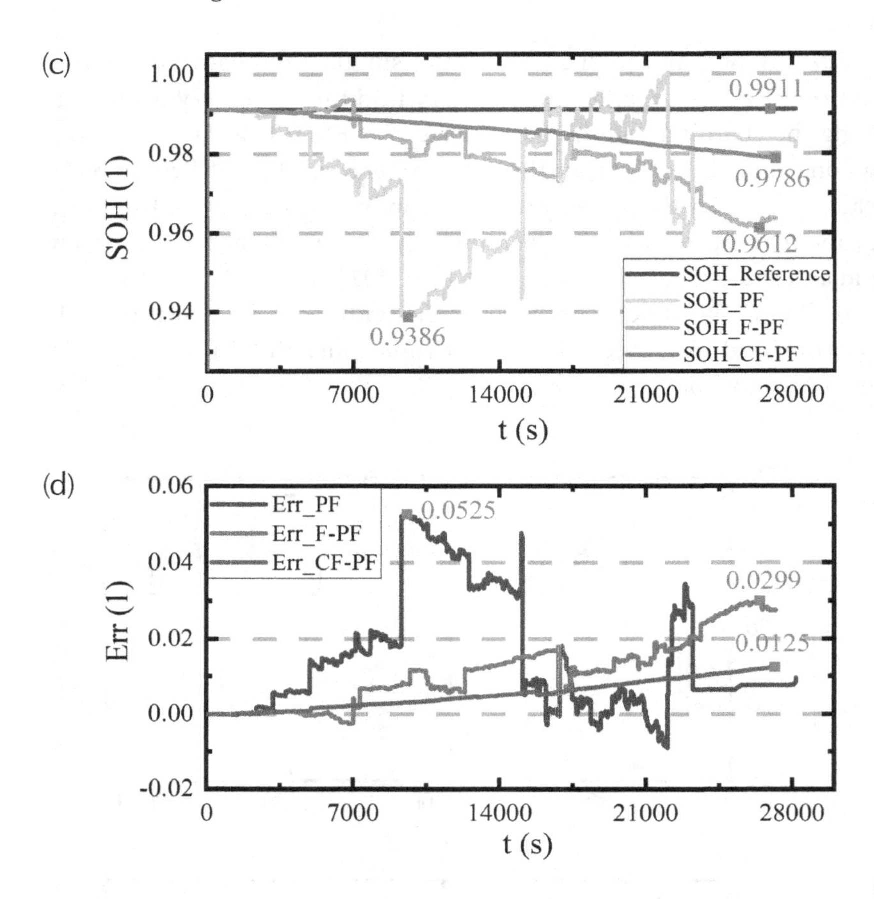


FIGURE 5.6 (Continued)

As can be seen from Figure 5.6(b), the SOC estimation error of the traditional PF algorithm fluctuates wildly, with high instability and a maximum error of up to 5.60%. The F-PF algorithm somewhat improved the overall estimation accuracy, with a maximum error of 4.16%. With the introduction of the chaotic algorithm, the overall performance of the CF-PF algorithm in SOC estimation has been dramatically improved, and the algorithm can maintain high stability even under complex and variable B BDST conditions. The algorithm can show convergence later on, with errors fluctuating around 0 and a maximum error of only 1.42%.

Figure 5.6(c) and Figure 5.6(d) show that the stability of the SOH estimation from the traditional PF algorithm is poor. In the middle stage of the discharge, the maximum error was up to 5.25%. The algorithm also showed an apparent instability in the late discharge period. Compared with the traditional PF algorithm, the F-PF algorithm has a significant improvement in overall stability and accuracy but still shows excellent divergence, with a maximum error of 2.99%. The improved CF-PF algorithm showed a good estimation effect in the whole BBDST condition, the error curve was non-stationary throughout the discharge period, and the algorithm has high stability, with a maximum error of only 1.25%. The three algorithms are compared by three evaluation indexes: ME, MAE, and RMSE. And Table 5.3 shows the comparison of synergistic estimation results of SOC and SOH under BBDST test conditions.

Figure 5.7 shows the comparison of the synergistic estimation results of SOC and SOH of the three algorithms, with ME, MAE, and RMSE as the evaluation indicators under the BBDST test conditions.

TABLE 5.3 Comparison of Collaborative Estimation Results of SOC and SOH under BB DST Conditions

| Estimation Method | PF | F-PF | CF-PF |
|---|---|---|---|
| ME (SOC) | 5.60% | 4.16% | 1.42% |
| MAE (SOC) | 2.48% | 1.31% | 0.52% |
| RMSE (SOC) | 2.76% | 1.54% | 0.67% |
| ME (SOH) | 5.25% | 2.99% | 1.25% |
| MAE (SOH) | 1.52% | 1.10% | 0.52% |
| RMSE (SOH) | 2.16% | 1.40% | 0.63% |

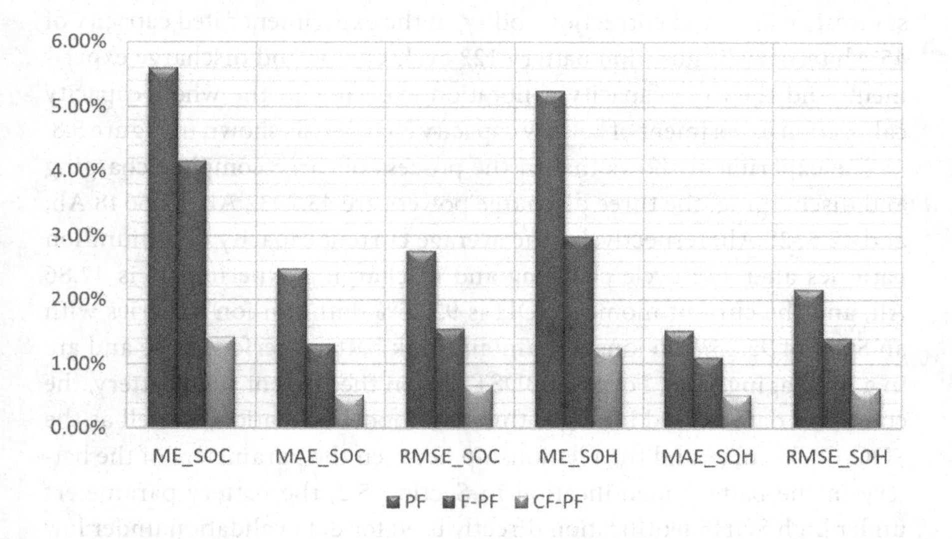


FIGURE 5.7 BBDST comparison of SOC and SOH under working conditions.

From the overall comparison results shown in Table 5.3 and Figure 5.7, the proposed CF-PF algorithm shows a noticeable improvement effect in SOC and SOH estimation, and the RMSE of the SOC estimation error is 0.67%, which is 2.09% better than the traditional PF algorithm; the RMSE of the mistake of SOH estimation is 0.63% and 1.53% better compared with the conventional PF algorithm. The effect of the CF-PF algorithm is improving accuracy and stability in BBDST condition.

## 5.3 CO-ESTIMATION VERIFICATION UNDER BATTERY AGING STATE

The migration model is constructed to adapt to the different aging states of lithium-ion batteries in actual use. The experimental verification done in the last section under other working conditions is all in the form of battery SOH of 99.11%, that is, in a high health state. To verify the adaptive adjustment ability of the dynamic migration model constructed in this study for different battery aging conditions and the adaptability of the proposed algorithm in the battery aging state, the test data of the battery condition under low SOH and the battery parameters identified under high SOH are directly used for data verification under low SOH.

### 5.3.1 Co-Estimation Verification under Mild Aging Condition

The previous experiment verification is in the high health battery condition; to verify the model and algorithm in the serious aging of battery state estimation and correction ability, in the experiment rated capacity of 45 Ah ternary lithium-ion battery 122 cycle charge and discharge experiment, and then the capacity calibration experiment, the whole capacity calibration experiment of battery capacity changes, as shown in Figure 5.8.

The experiment shows that in the process of three complete charging and discharging, the three discharge powers are 43.2932 Ah, 42.6648 Ah, and 42.6328 Ah, respectively. The average current capacity of lithium-ion batteries after 122 cycle charging and discharging experiments is 42.86 Ah, and the current moment SOH is 95.25%. Lithium-ion batteries with an SOH of 95.25% no longer maintain peak battery performance and are in a mild aging state. For the BBDST test on the current state battery, the condition data is used for algorithm and model validation, as well as the SOC-OCV curve and the relationship between the parameters of the battery in the battery identification in Section 5.2, the battery parameters under high SOH identification directly used for data validation under low

SOH, to verify the adaptive regulation function of the battery dynamic migration model. The cooperative estimation results of SOC and SOH between the traditional PF algorithm, the F-PF algorithm without the chaotic mapping algorithm, and the improved CF-PF algorithm when the SOH is 95.25% are shown in Figure 5.9.

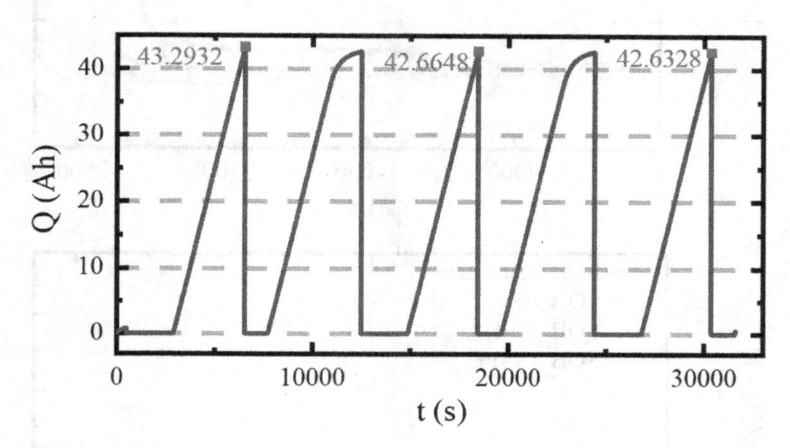


FIGURE 5.8 122 battery capacity calibration experiment after cycles of charge and discharge.

(a)

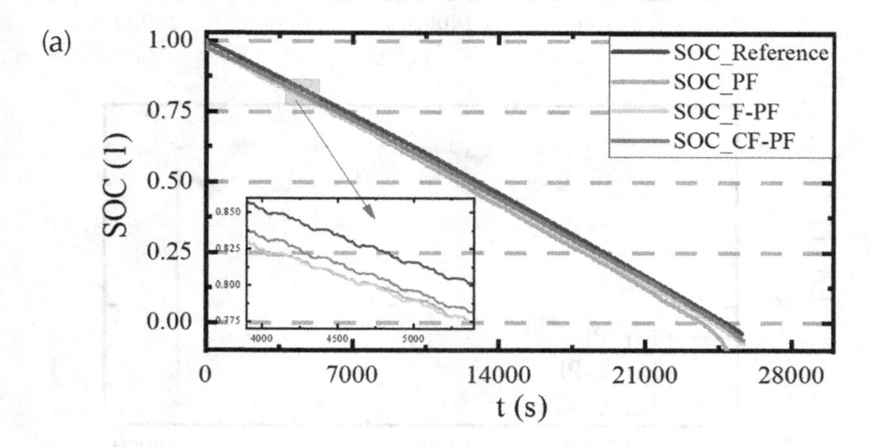


FIGURE 5.9 Comparison of cooperative estimation results of SOC and SOH in BBDST condition: (a) SOC estimation results of BBDST condition of a mild aging battery; (b) SOC estimation error of mild aged battery under BBDST condition; (c) SOH estimation results of mild aging battery under BBDST condition; (d) SOH estimation error of mild aging battery under BBDST condition.

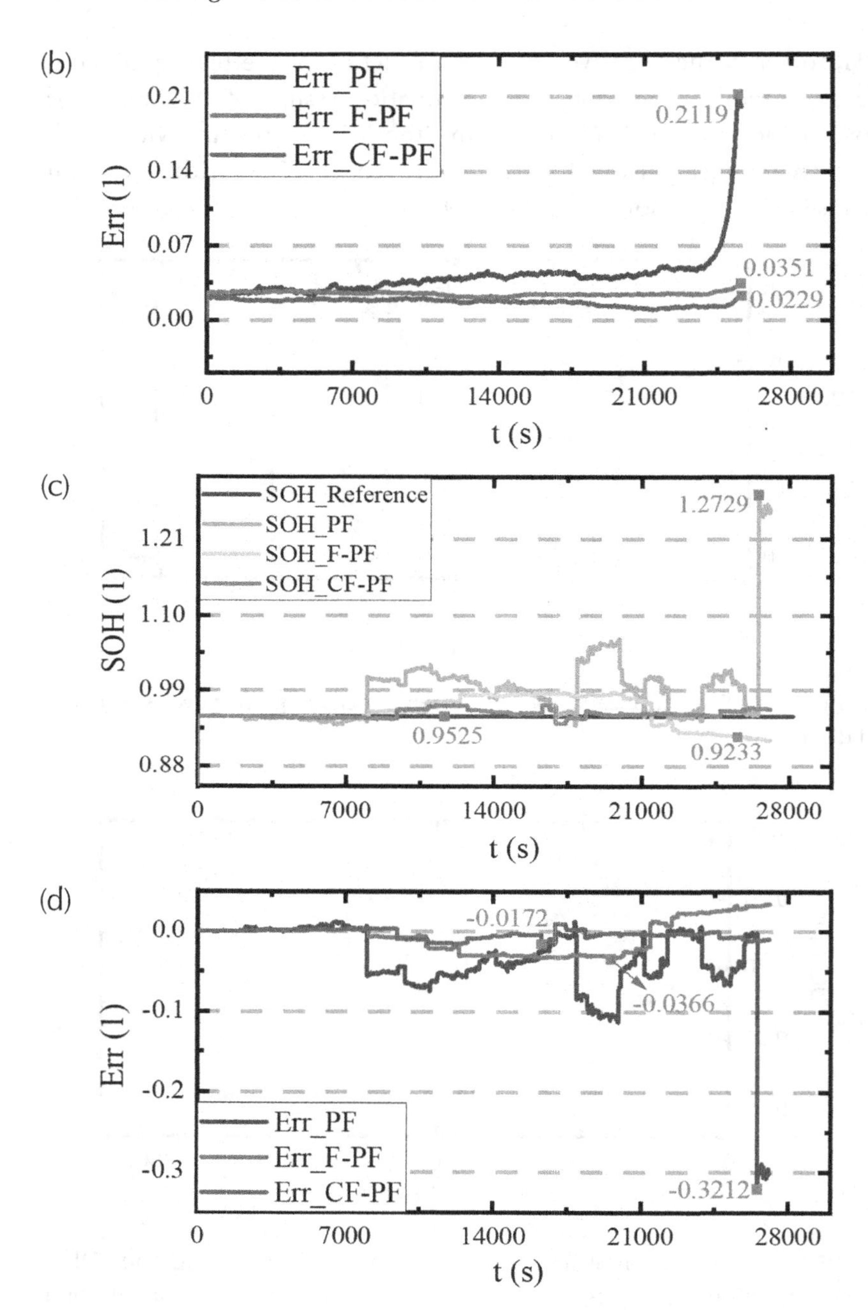


FIGURE 5.9 (Continued)

As can be seen from Figure 5.9(b), the traditional PF algorithm SOC estimation error is relatively stable, late has reached an uncontrollable divergence state, and the maximum error is as high as 21.19%, far beyond the acceptable, reasonable range. This shows that the battery dynamic migration model and traditional PF algorithm for mild aging state lithium-ion battery SOC estimation is not applicable due to a lack of error correction ability. However, the estimation results of the F-PF and CF-PF algorithms for SOC always maintain high stability throughout the discharge cycle, slightly diverging in the later period. Still, the error of the F-PF algorithm is only 3.51%, and the error of the CF-PF algorithm is only 2.29%, which is entirely within the acceptable, reasonable range.

As can be seen from Figure 5.9(c) and Figure 5.9(d), the SOH estimation effect of the traditional PF algorithm is volatile, the late algorithm diverges strongly, and the error is sharply increased to 32.12%. At this time, the SOH estimate has lost credibility. In contrast, the SOH estimation effect of the F-PF algorithm showed a vast improvement overall, with a maximum error of 3.66%. However, the CF-PF algorithm showed promising results in SOH estimation, and the overall error was stable and consistently tracked the actual value well, with no divergence in the later stage, and the maximum error was only 1.72%. The three algorithms are compared by ME, MAE, and RMSE, and Table 5.4 shows the comparison of SOC and SOH in BBDST test workers and mild aging.

Figure 5.10 shows the comparison of the synergistic estimation results of SOC and SOH of the three algorithms, with ME, MAE, and RMSE as the evaluation indicators under the BBDST mild aging test condition.

From the overall comparison results shown in Table 5.4 and Figure 5.10, the proposed CF-PF algorithm shows a noticeable improvement effect in SOC and SOH estimation. In the BBDST and mild aging (SOH = 95.25%), the RMSE of SOC estimation error is 1.24%, which is 1.99%

TABLE 5.4 Comparison of Synergistic Estimation Results of SOC and SOH for BBDST Mild Aging Battery (SOH = 95.25%)

| **Estimation Method** | **PF** | **F-PF** | **CF-PF** |
|---|---|---|---|
| ME (SOC) | 21.19% | 3.51% | 2.29% |
| MAE (SOC) | 2.98% | 1.97% | 1.23% |
| RMSE (SOC) | 3.23% | 1.98% | 1.24% |
| ME (SOH) | 32.12% | 2.99% | 1.25% |
| MAE (SOH) | 3.81% | 1.74% | 0.50% |
| RMSE (SOH) | 6.16% | 2.16% | 0.66% |

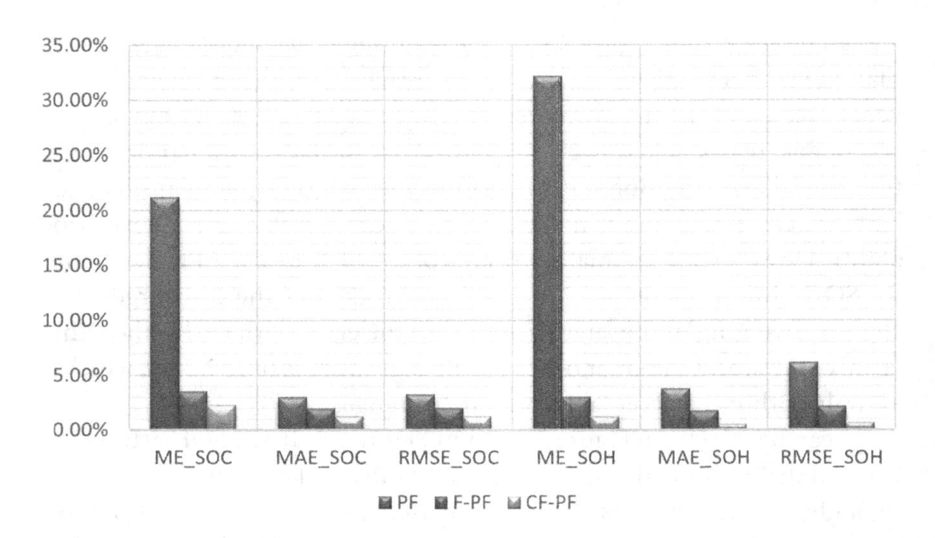


FIGURE 5.10 Comparison of SOC and SOH synergy estimated results for mild aging battery (SOH = 95.25%).

better than the traditional PF algorithm; the RMSE of SOH estimation error is 0.66% and 5.50% better compared with the conventional PF algorithm. In the case of BBDST and mild aging (SOH = 95.25%), the effect of the CF-PF algorithm has been verified for improving accuracy and stability in SOC and SOH co-estimation and proved that the dynamic migration model has a solid adaptive adjustment ability for battery aging.

### 5.3.2 Co-Estimation Verification under Severe Aging Condition

To further verify the battery migration model and algorithm in serious aging of the battery state estimation and correction ability, in the rated capacity of 45 Ah of a ternary lithium-ion battery 278 cycle charge and discharge experiment, then to the capacity calibration experiment, the whole capacity calibration experiment of battery capacity changes, as shown in Figure 5.11.

The experiment shows that in the process of three complete charging and discharging, the three discharge powers are 41.2211 Ah, 41.0450 Ah, and 41.0024 Ah, respectively. On average, the current capacity of the lithium-ion battery after 278 cycle charging and discharging experiments is 41.08 Ah, and the SOH at the current moment can be obtained as 91.31%. The performance of lithium-ion batteries with an SOH of

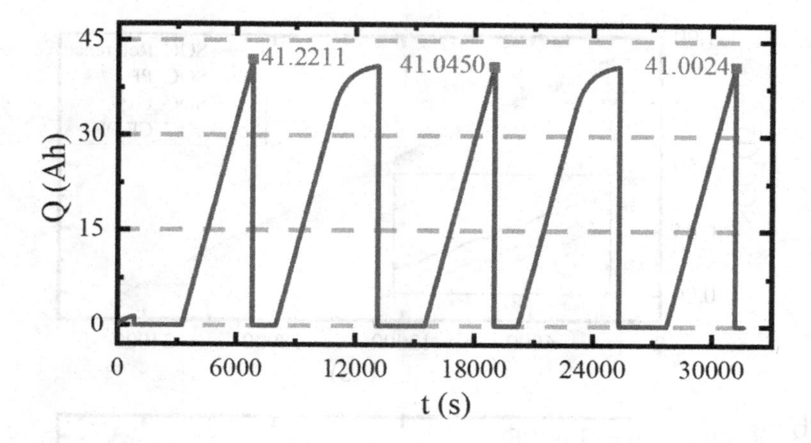


FIGURE 5.11 Battery capacity calibration experiment after 278 cycles of charge and discharge.

91.31% has been significantly reduced in the actual use process, mainly manifested by the reduction of battery durability and the need for frequent charging. When the battery SOH is reduced to 80%, it is considered to reach the failure threshold, cannot be used commonly, and needs to be scrapped. Therefore, the lithium-ion battery with an SOH of 91.31% is in a severe aging state. For the BBDST test on the current state battery, the condition data is used for algorithm and model validation, as well as the SOC-OCV curve and the relationship between the parameters of the battery in the battery identification in Section 5.2, the battery parameters under high SOH identification directly used for data validation under low SOH, to verify the adaptive regulation function of the battery dynamic migration model. The cooperative estimation results of SOC and SOH in the traditional PF algorithm, F-PF algorithm, and improved CF-PF algorithm when the SOH is 91.31% are shown in Figures 5.12.

As can be seen from Figure 5.12(b), although these three algorithms show high stability in the initial stage of discharge, the error of SOC estimated by the traditional PF algorithm at the end stage of liberation is seriously divergent, with a sharp increase in error, up to 21.24%, which is no longer within the acceptable reasonable range. The estimation error of the F-PF and CF-PF algorithms still showed high stability at the end of the discharge, compared with the relatively high overall accuracy of the CF-PF algorithm, with a maximum error of 3.46% for the F-PF algorithm and only 2.31% for the CF-PF algorithm.

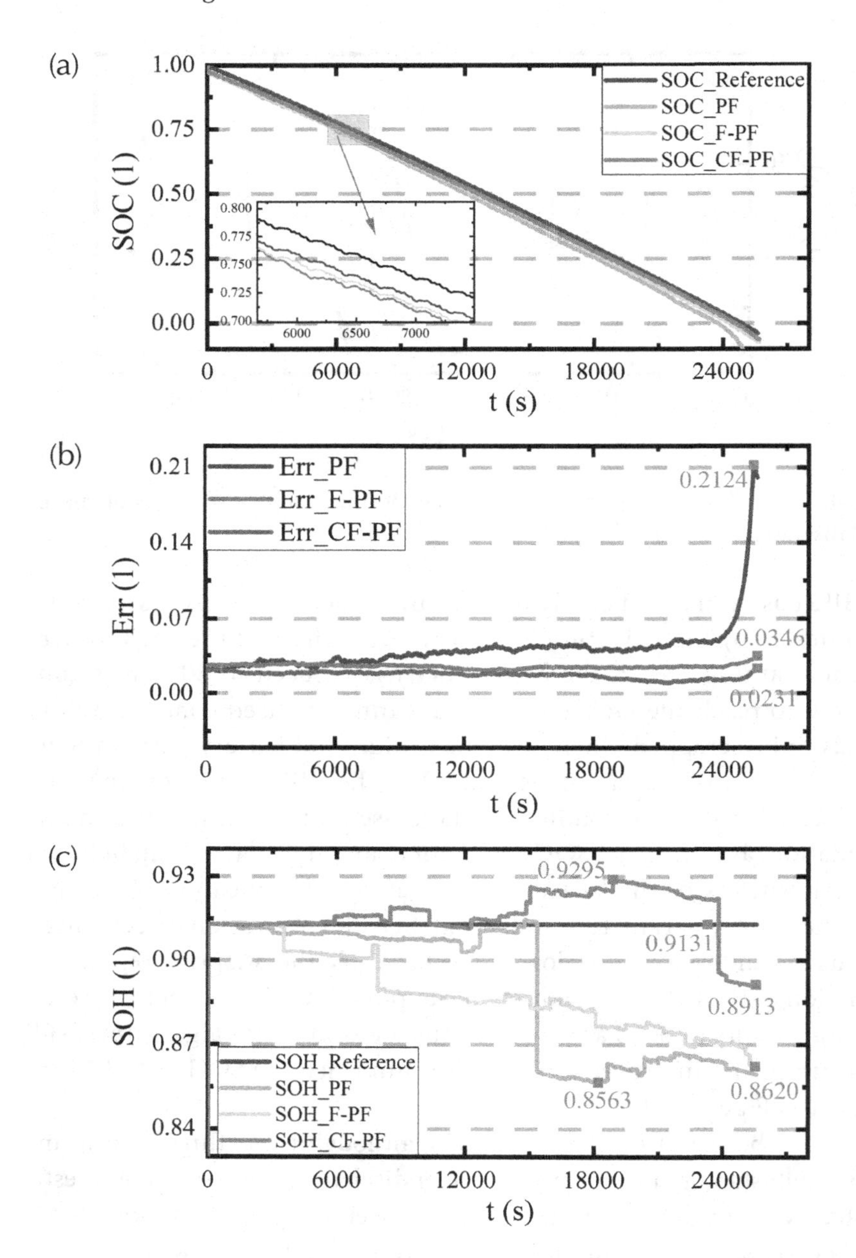


FIGURE 5.12 Comparison of SOC and SOH co-estimation results of severe aging battery under BBDST working condition: (a) SOC estimation results of severely aging battery under BBDST condition; (b) SOC estimation error of severely aging battery under BBDST condition; (c) SOH estimation results of severely aging battery under BBDST condition; (d) SOH estimation error of severely aging battery under BBDST condition.

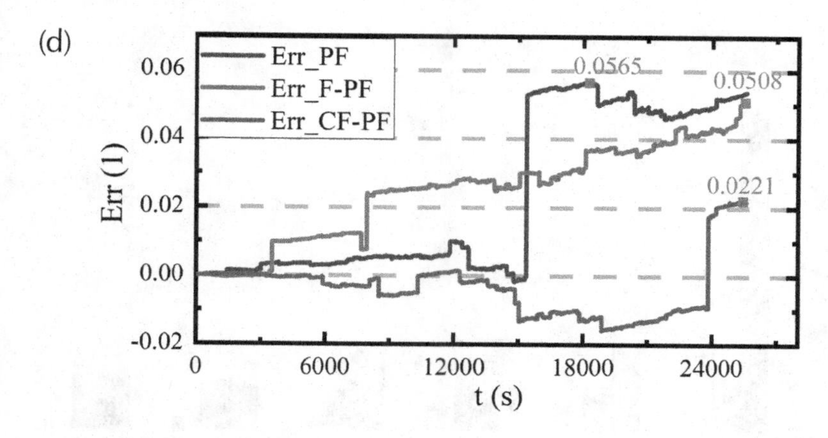


FIGURE 5.12 (Continued)

TABLE 5.5 Comparison of SOC and SOH (SOH = 91.31%)

| Estimation Method | PF | F-PF | CF-PF |
|---|---|---|---|
| ME (SOC) | 21.24% | 3.46% | 2.31% |
| MAE (SOC) | 5.76% | 3.08% | 2.09% |
| RMSE (SOC) | 6.03% | 3.13% | 2.18% |
| ME (SOH) | 5.65% | 5.08% | 2.21% |
| MAE (SOH) | 3.44% | 3.32% | 1.13% |
| RMSE (SOH) | 3.82% | 3.69% | 1.26% |

As can be seen from Figures 5.12(c) and Figures 5.12(d), the traditional PF algorithm has substantial errors in the SOH estimation, and the estimation results need to be more reliable. Although the SOH estimation effect of the F-PF algorithm is improved compared with PF, and although the maximum error is up to 5.08%, its overall accuracy is within the acceptable range. The results of the SOH estimation of the improved CF-PF algorithm have slight fluctuation and high precision, and the error can be controlled within 2.21%.

Although the overall effect of synergistic estimation in the severe aging state is relatively poor compared to the high SOH state, the CF-PF algorithm proposed in this study still has pronounced effects in improving the accuracy of synergistic estimation between SOC and SOH. The three algorithms are compared by ME, MAE, and RMSE, and Table 5.5 shows the comparison of SOC and SOH in BBDST test workers and severe aging.

Figure 5.13 shows the comparison of the synergistic estimation results of SOC and SOH of the three algorithms, with ME, MAE, and RMSE as the evaluation indicators under the BBDST severe aging test condition.

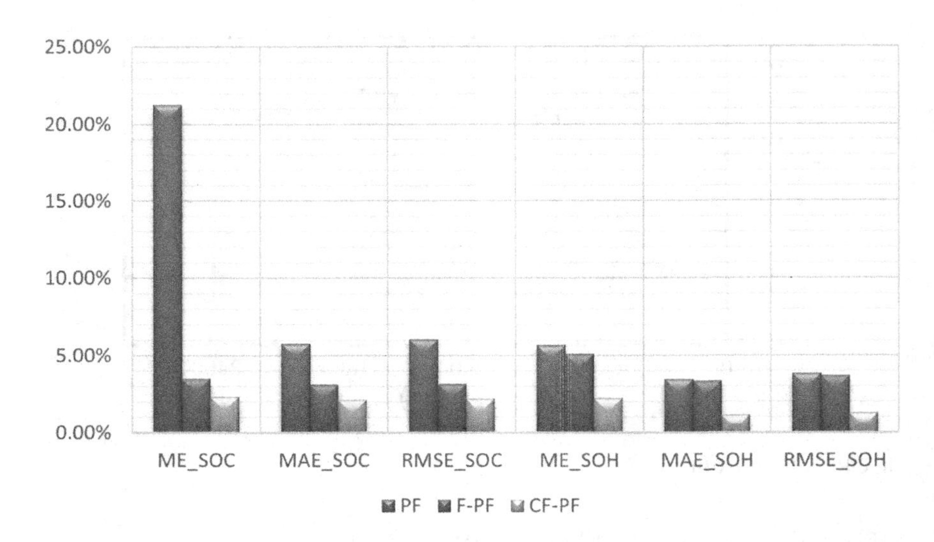


FIGURE 5.13 Comparison of SOC and SOH co-estimation for severe aging battery (SOH = 91.31%) under BBDST working condition.

According to the overall comparison results shown in Table 5.5 and Figure 5.13, the proposed CF-PF algorithm shows a noticeable improvement effect in SOC and SOH estimation. Under the BBDST and severe aging (SOH = 91.31%), the RMSE of the SOC estimation error is 2.18%, which is 3.85% better compared with the traditional PF algorithm; the RMSE of the SOH estimation error is 1.26% and 2.56% improvement compared with the conventional PF algorithm. In the case of BBDST and mild aging (SOH = 91.31%), the effect of the CF-PF algorithm on improving the accuracy and stability in the collaborative estimation of SOC and SOH was verified, and it was proved that the dynamic migration model still has a solid adaptive regulation ability for the severe aging of batteries.

## 5.4 SUMMARY OF THIS CHAPTER

In this chapter, based on the application of the dynamic migration battery model, the experimental condition of the CF-PF method under high health conditions (SOH = 99.11%), and the experimental results of the CF-PF method, the effectiveness of the proposed algorithm is proven. To verify the adaptive regulation effect of the migration model on the battery aging state, this chapter demonstrated the data of two different health conditions in mild aging (SOH = 95.25%) and severe aging (SOH = 91.31%), and the results are compared and analyzed to prove the effectiveness of dynamic battery migration model and CF-PF method.

CHAPTER 6

# Work Summary and Outlook

## 6.1 SUMMARY OF THE FULL TEXT OF THE WORK

As the technical bottleneck of the development of pure electric new energy vehicles, the performance level of power battery directly affects the use performance of pure electric vehicles. Two key states that must be accurately evaluated in a battery management system, SOC and SOH, are the necessary prerequisites for realizing balance control, charge and discharge strategy adjustment, fault diagnosis, and other functions. To realize the accurate evaluation of power battery SOC and SOH, it is of great significance to improve the overall performance of the battery in the whole life cycle. Therefore, this book takes ternary lithium-ion battery as the research object and carries out the collaborative estimation study of lithium-ion battery SOC and SOH based on the dynamic migration model and firefly optimization algorithm, mainly completing the following research contents:

1. After obtaining the current and voltage data of HPPC, DST, and BBDST by formulating and conducting experiments, the functional relationship between the open circuit voltage and SOC was obtained by analyzing and extracting the data and curve fitting. The charge and discharge characteristics and capacity attenuation characteristics were studied, the influence of discharge rate on SOC change, the influence of capacity attenuation on battery internal parameters and

DOI: 10.1201/9781003486251-6

SOH were analyzed, and the relationship between aging characteristics and SOH change was studied.

2. After establishing the initial offline Thevenin model, the functional relationship between each parameter and the SOC of the battery is obtained through offline parameter identification. The migration factor is added to the functional relationship to realize the adjustment of each parameter. The migration factor forms the migration matrix, and the particle filter algorithm is used to continuously correct and update the migration matrix to realize the dynamic adjustment of the battery parameters and then realize the adaptive adjustment of the built model according to the aging situation of different batteries.

3. Realize progressive cycle correction between SOC and SOH. The first-layer particle filter realizes the SOC estimation, and the estimated value is used as the input of the second-layer filter to realize the progressive estimation of SOH. Then, the current moment SOH estimate obtained in the second-layer Kalman filter is taken as input for the next iteration cycle, and the SOC estimate of the next moment is further corrected to form a closed loop until the end of the iteration. Then, the firefly algorithm is applied in the optimization process of SOC and SOH particles to realize the proximity of the particle to the optimal value, that is, the state particle to the actual value. On this basis, the chaos mapping algorithm is added to linearly map the variables into chaotic variables and then optimize the search process according to the ergodicity and randomness of chaos to achieve the high-precision collaborative estimation of SOC and SOH.

4. To verify the effectiveness of the dynamic migration model and optimization of the population firefly algorithm through three complex test conditions of SOC and SOH collaborative estimation of the experiment and in the battery health, battery mild aging, and battery severe aging validation model and algorithm, and the traditional algorithm under the estimation results of the comparison, the detailed analysis of the estimation results. The experimental results in the battery health state show that the RMSE of the SOC estimation error of the CF-PF algorithm is 1.48%, which is a 1.00% improvement compared with the traditional PF algorithm; the RMSE of the SOH estimation error is 0.38%, which is 0.94% improvement compared with the conventional PF algorithm. The RMSE of the SOC estimation error of the CF-PF algorithm

under the DST condition is 2.70%, which improves 0.41% compared to the traditional PF algorithm; the RMSE of the SOH estimation error is 0.17%, which enhances 0.52% compared to the conventional PF algorithm. The RMSE of the SOC estimation error of the CF-PF algorithm is 0.67%, which is 2.09% better compared with the traditional PF algorithm; the RMSE of the SOH estimation error is 0.63%, which is 1.53% better compared with the conventional PF algorithm. In BBDST and mild aging (SOH = 95.25%), the RMSE of its SOC estimation error is 1.24%, 1.99% better than the traditional PF algorithm; the RMSE of its SOH estimation error is 0.66%, 5.50% better than the conventional PF algorithm. In BBDST and severe aging (SOH = 91.31%), the RMSE of the SOC estimation error is 2.18%, which is 3.85% better than the traditional PF algorithm; the RMSE of the SOH estimation error is 1.26%, 2.56% better than the conventional PF algorithm. The preceding experimental results and comparative analysis verify the effectiveness of the proposed model and method of this study.

Battery health condition and the experimental validation results show that the dynamic battery migration model and chaotic firefly-particle filtering method can effectively improve the dynamic lithium-ion battery SOC and SOH estimated accuracy and estimated stability and provide a theoretical basis for the effective management and safety of the power lithium-ion battery.

## 6.2 OUTLOOK FOR FUTURE RESEARCH

This study aims to achieve a high-precision collaborative estimation of the charge state and health state of powered lithium-ion batteries. The researchers conducted a series of studies on experimental design, algorithm optimization, and experimental verification. These studies have effectively improved the accuracy and convergence of the co-estimation of the state of charge and state of health and have made some progress. The results of this research have certain practical significance for the effective management and safety application of lithium-ion batteries in new energy electric vehicles, but there are still certain shortcomings:

1. Although this study considers the impact of battery aging on SOC and SOH estimation caused by extended battery use and repeated charging and discharge in actual use, the effect of temperature on

the battery is not considered. In practical application, it is inevitable to encounter highly high temperatures and bad weather, and the impact of environmental temperature on the performance of lithium-ion batteries is very significant. Therefore, in future studies, temperature will be added as an influencing factor in lithium-ion battery modeling to build a more accurate dynamic migration model.

2. Although this study has estimated the SOC and SOH of lithium-ion battery cells considering aging, in the actual use of new energy electric vehicles, the battery cell cannot support the operation of huge vehicles and often appears in the form of lithium-ion battery packs. Consider the state estimation of lithium-ion battery packs, which will significantly increase the difficulty and complexity of the research. Therefore, the author will improve the experimental design and development of the lithium-ion battery pack in the subsequent research work and consider the battery pack balance.

# References

Al-Gabalawy, Mostafa, Nesreen S. Hosny, James A. Dawson, and Ahmed I. Omar. 2021. "State of Charge Estimation of a Li-Ion Battery Based on Extended Kalman Filtering and Sensor Bias." *International Journal of Energy Research* 45 (5): 6708–6726. https://doi.org/10.1002/er.6265. <Go to ISI>://WOS:000596468300001.

Alhadawi, Hussam S., Dragan Lambic, Mohamad Fadli Zolkipli, and Musheer Ahmad. 2020. "Globalized Firefly Algorithm and Chaos for Designing Substitution Box." *Journal of Information Security and Applications* 55. https://doi.org/10.1016/j.jisa.2020.102671. <Go to ISI>://WOS:000601366700005.

Ali, Muhammad U., Hafiz F. Khan, Haris Masood, Karam D. Kallu, Malik M. Ibrahim, Amad Zafar, Semin Oh, and Sangil Kim. 2022. "An Adaptive State of Charge Estimator for Lithium-Ion Batteries." *Energy Science & Engineering* 10 (7): 2333–2347. https://doi.org/10.1002/ese3.1141. <Go to ISI>://WOS:000776531000001.

Ali, Omer, Mohamad Khairi Ishak, Ashraf Bani Ahmed, Mohd Fadzli Mohd Salleh, Chia Ai Ooi, Muhammad Firdaus Akbar Jalaludin Khan, and Imran Khan. 2022. "On-Line WSN SoC Estimation Using Gaussian Process Regression: An Adaptive Machine Learning Approach." *Alexandria Engineering Journal* 61 (12): 9831–9848. https://doi.org/10.1016/j.aej.2022.02.0671110-0168. <Go to ISI>://WOS:000806179500013.

Alqahtani, Abdullah, Shtwai Alsubai, Adel Binbusayyis, Mohemmed Sha, Abdu Gumaei, and Yu-Dong Zhang. 2023. "Prediction of Urinary Tract Infection in IoT-Fog Environment for Smart Toilets Using Modified Attention-Based ANN and Machine Learning Algorithms." *Applied Sciences-Basel* 13 (10). https://doi.org/10.3390/app13105860. <Go to ISI>://WOS:000994274100001.

Anand, Priyanka, and Sankalap Arora. 2020. "A Novel Chaotic Selfish Herd Optimizer for Global Optimization and Feature Selection." *Artificial Intelligence Review* 53 (2): 1441–1486. https://doi.org/10.1007/s10462-019-09707-6. <Go to ISI>://WOS:000513275100020.

Arora, Sankalap, Manik Sharma, and Priyanka Anand. 2020. "A Novel Chaotic Interior Search Algorithm for Global Optimization and Feature Selection." *Applied Artificial Intelligence* 34 (4): 292–328. https://doi.org/10.1080/08839514.2020.1712788. <Go to ISI>://WOS:000508818700001.

Askari, Iman, Mulugeta A. Haile, Xuemin Tu, and Huazhen Fang. 2022. "Implicit Particle Filtering Via a Bank of Nonlinear Kalman Filters?" *Automatica* 145. https://doi.org/10.1016/j.automatica.2022.110469. <Go to ISI>://WOS:000859030000005.

Atali, Gokhan, Ihsan PehlIvan, Bilal Gurevin, and Halil Ibrahim Seker. 2021. "Chaos in Metaheuristic Based Artificial Intelligence Algorithms: A Short Review." *Turkish Journal of Electrical Engineering and Computer Sciences* 29 (3): 1354–1367. https://doi.org/10.3906/elk-2102-5. <Go to ISI>://WOS:000679315700003.

Aydilek, Ibrahim Berkan, Izzettin Hakan Karacizmeli, Mehmet Emin Tenekeci, Serkan Kaya, and Abdulkadir Gumuscu. 2021. "Using Chaos Enhanced Hybrid Firefly Particle Swarm Optimization Algorithm for Solving Continuous Optimization Problems." *Sadhana-Academy Proceedings in Engineering Sciences* 46 (2). https://doi.org/10.1007/s12046-021-01572-w. <Go to ISI>://WOS:000636289900006.

Aylagas, Raul Ciria, Clara Ganuza, Ruben Parra, Maria Yanez, and Elixabete Ayerbe. 2022. "cideMOD: An Open Source Tool for Battery Cell Inhomogeneous Performance Understanding." *Journal of the Electrochemical Society* 169 (9). https://doi.org/10.1149/1945-7111/ac91fb. <Go to ISI>://WOS:000860234500001.

Babaeiyazdi, Iman, Afshin Rezaei-Zare, and Shahab Shokrzadeh. 2021. "State of Charge Prediction of EV Li-Ion Batteries Using EIS: A Machine Learning Approach." *Energy* 223. https://doi.org/10.1016/j.energy.2021.120116. <Go to ISI>://WOS:000637964000004.

Babu, Anandh Ramesh. 2022. *Battery Thermal Management for Electric Vehicles Operating in Cold Climates*. Edited by Anandh Ramesh Babu. Ann Arbor: ProQuest Dissertations Publishing.

Bai, Wenyuan, Xinhui Zhang, Zhen Gao, Shuyu Xie, Yu Chen, Yu He, and Jun Zhang. 2022. "State of Charge Estimation for Lithium-Ion Batteries Under Varying Temperature Conditions Based on Adaptive Dual Extended Kalman Filter." *Electric Power Systems Research* 213. https://doi.org/10.1016/j.epsr.2022.108751. <Go to ISI>://WOS:000863067700001.

Bai, Xiangtao, Liqing Ban, and Weidong Zhuang. 2020. "Research Progress on Coating and Doping Modification of Nickel Rich Ternary Cathode Materials." *Journal of Inorganic Materials* 35 (9): 972–986. https://doi.org/10.15541/jim20190568. <Go to ISI>://WOS:000562923400012.

Bao, Yun, and Yinchu Gong. 2023. "Li-Ion Battery Charge Transfer Stability Studies with Direct Current Impedance Spectroscopy." *Energy Reports* 9: 34–41. https://doi.org/10.1016/j.egyr.2023.03.002. <Go to ISI>://WOS:001057956800004.

Barai, Anup, W. Dhammika Widanage, James Marco, Andrew McGordon, and Paul Jennings. 2015. "A Study of the Open Circuit Voltage Characterization Technique and Hysteresis Assessment of Lithium-Ion Cells." *Journal of Power Sources* 295: 99–107. https://doi.org/10.1016/j.jpowsour.2015.06.140. <Go to ISI>://WOS:000359330500013.

Bavand, Amin, S. Ali Khajehoddin, Masoud Ardakani, and Ahmadreza Tabesh. 2022. "Online Estimations of Li-Ion Battery SOC and SOH Applicable to Partial Charge/Discharge." *IEEE Transactions on Transportation Electrification* 8 (3): 3673–3685. https://doi.org/10.1109/tte.2022.3162164. <Go to ISI>://WOS:000837770500052.

Bayatinejad, M. A., and A. Mohammadi. 2021. "Investigating the Effects of Tabs Geometry and Current Collectors Thickness of Lithium-Ion Battery with Electrochemical-Thermal Simulation." *Journal of Energy Storage* 43. https://doi.org/10.1016/j.est.2021.103203. <Go to ISI>://WOS:000701775700005.

Behera, Mandakini, Archana Sarangi, Debahuti Mishra, Pradeep Kumar Mallick, Jana Shafi, Parvathaneni Naga Srinivasu, and Muhammad Fazal Ijaz. 2022. "Automatic Data Clustering by Hybrid Enhanced Firefly and Particle Swarm Optimization Algorithms." *Mathematics* 10 (19). https://doi.org/10.3390/math10193532. <Go to ISI>://WOS:000868011200001.

Bian, Xiaolei, Zhongbao Wei, Jiangtao He, Fengjun Yan, and Longcheng Liu. 2021. "A Novel Model-Based Voltage Construction Method for Robust State-of-Health Estimation of Lithium-Ion Batteries." *IEEE Transactions on Industrial Electronics* 68 (12): 12173–12184. https://doi.org/10.1109/tie.2020.3044779. <Go to ISI>://WOS:000692884200053.

Bian, Xiaolei, Zhongbao Gae Wei, Weihan Li, Josep Pou, Dirk Uwe Sauer, and Longcheng Liu. 2022. "State-of-Health Estimation of Lithium-Ion Batteries by Fusing an Open Circuit Voltage Model and Incremental Capacity Analysis." *IEEE Transactions on Power Electronics* 37 (2): 2226–2236. https://doi.org/10.1109/tpel.2021.3104723. <Go to ISI>://WOS:000707555600086.

Bobobee, Etse Dablu, Shunli Wang, Chuanyun Zou, Emmanuel Appiah, Heng Zhou, Paul Takyi-Aninakwa, and Md Amdadul Haque. 2022. "Improved Fixed Range Forgetting Factor-Adaptive Extended Kalman Filtering (FRFF-AEKF) Algorithm for the State of Charge Estimation of High-Power Lithium-Ion Batteries." *International Journal of Electrochemical Science* 17 (11). https://doi.org/10.20964/2022.11.46. <Go to ISI>://WOS:000968974300017.

Bobobee, Etse Dablu, Shunli Wang, Chuanyun Zou, Paul Takyi-Aninakwa, Heng Zhou, and Emmanuel Appiah. 2023. "State of Charge Estimation of Ternary Lithium-Ion Batteries at Variable Ambient Temperatures." *International Journal of Electrochemical Science* 18 (4). https://doi.org/10.1016/j.ijoes.2023.100062. <Go to ISI>://WOS:000950072600001.

Boerger, Elisabeth Maria, Felix Gottschalk, Laura Drescher, and Alexander Boerger. 2022. "Shaking Lithium-Ion Cells on a Rocker – Temperature-Dependent Influence of Mechanical Movement on Lithium-Ion Battery Lifetime." *Chemie Ingenieur Technik* 94 (4): 603–606. https://doi.org/10.1002/cite.201800218. <Go to ISI>://WOS:000675260100001.

Braco, Elisa, Idoia San Martin, Pablo Sanchis, Alfredo Ursua, and Daniel-Ioan Stroe. 2022. "State of Health Estimation of Second-Life Lithium-Ion Batteries Under Real Profile Operation." *Applied Energy* 326. https://doi.org/10.1016/j.apenergy.2022.119992. <Go to ISI>://WOS:000862876900005.

Brucker, Jennifer, Rene Behmann, Wolfgang G. Bessler, and Rainer Gasper. 2022. "Neural Ordinary Differential Equations for Grey-Box Modelling of Lithium-Ion Batteries on the Basis of an Equivalent Circuit Model." *Energies* 15 (7). https://doi.org/10.3390/en15072661. <Go to ISI>://WOS:000781366900001.

Bustos, Richard, S. Andrew Gadsden, Mohammad Al-Shabi, and Shohel Mahmud. 2023. "Lithium-Ion Battery Health Estimation Using an Adaptive Dual Interacting Model Algorithm for Electric Vehicles." *Applied Sciences-Basel* 13 (2). https://doi.org/10.3390/app13021132. <Go to ISI>://WOS:000916922500001.

Cai, Kedi, Ya Zhou, Shuang Yan, Lan Li, and Xiaoshi Lang. 2023. "Construction of Oriented-Rod Structure Ni, Mn Co-Doped Vanadium Oxide Ternary Materials Synthesized by a Facile Coprecipitation Method as High-Performance Cathode Active Material for Lithium-Ion Capacitive-Batteries." *Ionics* 29 (2): 573–579. https://doi.org/10.1007/s11581-022-04853-4. <Go to ISI>://WOS:000898286700001.

Cen, Zhaohui, and Pierre Kubiak. 2020. "Lithium-Ion Battery SOC/SOH Adaptive Estimation Via Simplified Single Particle Model." *International Journal of Energy Research* 44 (15): 12444–12459. https://doi.org/10.1002/er.5374. <Go to ISI>://WOS:000523296600001.

Chang, Chun, Qiyue Wang, Jiuchun Jiang, Yan Jiang, and Tiezhou Wu. 2023. "Voltage Fault Diagnosis of a Power Battery Based on Wavelet Time-Frequency Diagram." *Energy* 278. https://doi.org/10.1016/j.energy.2023.127920. <Go to ISI>://WOS:001021676200001.

Chang, Chun, Qiyue Wang, Jiuchun Jiang, and Tiezhou Wu. 2021. "Lithium-Ion Battery State of Health Estimation Using the Incremental Capacity and Wavelet Neural Networks with Genetic Algorithm." *Journal of Energy Storage* 38. https://doi.org/10.1016/j.est.2021.102570. <Go to ISI>://WOS:000663165300009.

Chang, Gi Hwan, Han Ul Choi, Sung Kang, Jun-Young Park, and Hyung-Tae Lim. 2020. "Characterization of Limiting Factors of an All-Solid-State Li-Ion Battery Using an Embedded Indium Reference Electrode." *Ionics* 26 (3): 1555–1561. https://doi.org/10.1007/s11581-019-03367-w. <Go to ISI>://WOS:000504170000003.

Chang, Xin, Yu-Ming Zhao, Boheng Yuan, Min Fan, Qinghai Meng, Yu-Guo Guo, and Li-Jun Wan. 2023. "Solid-State Lithium-Ion Batteries for Grid Energy Storage: Opportunities and Challenges." *Science China-Chemistry*. https://doi.org/10.1007/s11426-022-1525-3. <Go to ISI>://WOS:000935049200001.

Che, Yanbo, Yibin Cai, Hongfeng Li, Yushu Liu, Mingda Jiang, and Peijun Qin. 2022. "State of Health Estimation Method for Lithium-Ion Batteries Based on Nonlinear Autoregressive Neural Network Model With Exogenous Input." *Journal of Electrochemical Energy Conversion and Storage* 19 (2). https://doi.org/10.1115/1.4052274. <Go to ISI>://WOS:000778139700009.

Che, Yanbo, Yushu Liu, Ze Cheng, and Ji'ang Zhang. 2021. "SOC and SOH Identification Method of Li-Ion Battery Based on SWPSO-DRNN." *IEEE Journal of Emerging and Selected Topics in Power Electronics* 9 (4):

4050–4061. https://doi.org/10.1109/jestpe.2020.3004972. <Go to ISI>://WOS:000679548500023.

Che, Yunhong, Zhongwei Deng, Xiaolin Tang, Xianke Lin, Xianghong Nie, and Xiaosong Hu. 2022. "Lifetime and Aging Degradation Prognostics for Lithium-ion Battery Packs Based on a Cell to Pack Method." *Chinese Journal of Mechanical Engineering* 35 (1). https://doi.org/10.1186/s10033-021-00668-y. <Go to ISI>://WOS:000740586800001.

Chen, Dongdong, Long Xiao, Wenduan Yan, and Yinbiao Guo. 2021. "A Novel Hybrid Equivalent Circuit Model for Lithium-Ion Battery Considering Nonlinear Capacity Effects." *Energy Reports* 7: 320–329. https://doi.org/10.1016/j.egyr.2021.06.051. <Go to ISI>://WOS:000749275000020.

Chen, Haodeng, Jianxing Xu, Shaomin Ji, Wenjin Ji, Lifeng Cui, and Yanping Huo. 2020. "Application of MOFs Derived Metal Oxides and Composites in Anode Materials of Lithium Ion Batteries." *Progress in Chemistry* 32 (2–3): 298–308. https://doi.org/10.7536/pc190610. <Go to ISI>://WOS:000542989400013.

Chen, Jianlong, Chenghao Zhang, Cong Chen, Chenlei Lu, and Dongji Xuan. 2023. "State-of-Charge Estimation of Lithium-Ion Batteries Using Convolutional Neural Network With Self-Attention Mechanism." *Journal of Electrochemical Energy Conversion and Storage* 20 (3). https://doi.org/10.1115/1.4055985. <Go to ISI>://WOS:001021827200005.

Chen, Junxiong, Xiong Feng, Lin Jiang, and Qiao Zhu. 2021. "State of Charge Estimation of Lithium-Ion Battery Using Denoising Autoencoder and Gated Recurrent Unit Recurrent Neural Network." *Energy* 227. https://doi.org/10.1016/j.energy.2021.120451. <Go to ISI>://WOS:000652610900009.

Chen, Lu, Shunli Wang, Lei Chen, Jialu Qiao, and Carlos Fernandez. 2023. "High-Precision State of Charge Estimation of Lithium-Ion Batteries Based on Improved Particle Swarm Optimization-Backpropagation Neural Network-Dual Extended Kalman Filtering." *International Journal of Circuit Theory and Applications*. https://doi.org/10.1002/cta.3788. <Go to ISI>://WOS:001066020300001.

Chen, Quanwei, Xin Lai, Huanghui Gu, Xiaopeng Tang, Furong Gao, Xuebing Han, and Yuejiu Zheng. 2022. "Investigating Carbon Footprint and Carbon Reduction Potential Using a Cradle-to-Cradle LCA Approach on Lithium-Ion Batteries for Electric Vehicles in China." *Journal of Cleaner Production* 369. https://doi.org/10.1016/j.jclepro.2022.133342. <Go to ISI>://WOS:000858457900002.

Chen, Saihan, Jinlei Sun, Shengshi Qiu, Xinwei Liu, Kai Lyu, Siwen Chen, Shiyou Xing, and Yilong Guo. 2023. "Phased Control Reciprocating Airflow Cooling Strategy for a Battery Module Considering Stage of Charge and State of Health Inconsistency." *Journal of Energy Storage* 61. https://doi.org/10.1016/j.est.2023.106752. <Go to ISI>://WOS:000963175100001.

Chen, Shan, Tianhong Pan, and Bowen Jin. 2023. "State of Charge Estimation of Lithium-Ion Battery Using Energy Consumption Analysis." *International Journal of Automotive Technology* 24 (2): 445–457. https://doi.org/10.1007/s12239-023-0037-2. <Go to ISI>://WOS:000953162900012.

Chen, Yong, Changlong Li, Sizhong Chen, Hongbin Ren, and Zepeng Gao. 2021. "A Combined Robust Approach Based on Auto-Regressive Long Short-Term Memory Network and Moving Horizon Estimation for State-of-Charge Estimation of Lithium-Ion Batteries." *International Journal of Energy Research* 45 (9): 12838–12853. https://doi.org/10.1002/er.6615. <Go to ISI>://WOS:000634453200001.

Chen, Yuhong. 2022. "Recent Advances of Overcharge Investigation of Lithium-Ion Batteries." *Ionics* 28 (2): 495–514. https://doi.org/10.1007/s11581-021-04331-3. <Go to ISI>://WOS:000716848400001.

Cheng, Qinglin, Xue Wang, Shuang Wang, Yanting Li, Hegao Liu, Zhidong Li, and Wei Sun. 2023. "Research on a Carbon Emission Prediction Method for Oil Field Transfer Stations Based on an Improved Genetic Algorithm-the Decision Tree Algorithm." *Processes* 11 (9). https://doi.org/10.3390/pr11092738. <Go to ISI>://WOS:001074369100001.

Cheng, Zhiwen, Haohao Song, Debin Zheng, Meng Zhou, and Kexin Sun. 2023. "Hybrid Firefly Algorithm with a New Mechanism of Gender Distinguishing for Global Optimization." *Expert Systems with Applications* 224. https://doi.org/10.1016/j.eswa.2023.120027. <Go to ISI>://WOS:000980762200001.

Chou, Jui-Sheng, and Truong Dinh-Nhat. 2020. "Multiobjective Optimization Inspired by Behavior of Jellyfish for Solving Structural Design Problems." *Chaos Solitons & Fractals* 135. https://doi.org/10.1016/j.chaos.2020.109738. <Go to ISI>://WOS:000540074900007.

Couto, Luis D., Raffaele Romagnoli, Saehong Park, Dong Zhang, Scott J. Moura, Michel Kinnaert, and Emanuele Garone. 2022. "Faster and Healthier Charging of Lithium-Ion Batteries Via Constrained Feedback Control." *IEEE Transactions on Control Systems Technology* 30 (5): 1990–2001. https://doi.org/10.1109/tcst.2021.3135149. <Go to ISI>://WOS:000734073600001.

Cui, Zhenhua, Le Kang, Liwei Li, Licheng Wang, and Kai Wang. 2022a. "A Combined State-of-Charge Estimation Method for Lithium-Ion Battery Using an Improved BGRU Network and UKF." *Energy* 259: 256–268. https://doi.org/10.1016/j.energy.2022.124933. <Go to ISI>://WOS:000848270400005.

Cui, Zhenhua, Le Kang, Liwei Li, Licheng Wang, and Kai Wang. 2022b. "A Combined State-of-Charge Estimation Method for Lithium-Ion Battery Using an Improved BGRU Network and UKF." *Energy* 259. https://doi.org/10.1016/j.energy.2022.124933. <Go to ISI>://WOS:000848270400005.

Cui, Zhenhua, Licheng Wang, Qiang Li, and Kai Wang. 2022. "A Comprehensive Review on the State of Charge Estimation for Lithium-Ion Battery Based on Neural Network." *International Journal of Energy Research* 46 (5): 5423–5440. https://doi.org/10.1002/er.7545. <Go to ISI>://WOS:000730608200001.

Dai, Jindong, Chi Zhai, Jiali Ai, Guangren Yu, Haichao Lv, Wei Sun, and Yongzhong Liu. 2023. "A Cellular Automata Framework for Porous Electrode Reconstruction and Reaction-Diffusion Simulation." *Chinese Journal of Chemical Engineering* 60: 262–274. https://doi.org/10.1016/j.cjche.2023.01.022. <Go to ISI>://WOS:001051078900001.

Das, Swagato, and Purnachandra Saha. 2021. "Performance of Swarm Intelligence Based Chaotic Meta-Heuristic Algorithms in Civil Structural Health Monitoring." *Measurement* 169. https://doi.org/10.1016/j.measurement.2020.108533. <Go to ISI>://WOS:000600432400004.

Dash, Sujata, Ajith Abraham, Ashish Kr Luhach, Jolanta Mizera-Pietraszko, and Joel J. P. C. Rodrigues. 2020. "Hybrid Chaotic Firefly Decision Making Model for Parkinson's Disease Diagnosis." *International Journal of Distributed Sensor Networks* 16 (1). https://doi.org/10.1177/1550147719895210. <Go to ISI>://WOS:000507245300001.

Dechent, Philipp, Alexander Epp, Dominik Jost, Yuliya Preger, Peter M. Attia, Weihan Li, and Dirk Uwe Sauer. 2021. "ENPOLITE: Comparing Lithium-Ion Cells across Energy, Power, Lifetime, and Temperature." *ACS Energy Letters* 6 (6): 2351–2355. https://doi.org/10.1021/acsenergylett.1c00743. <Go to ISI>://WOS:000662227100040.

Degen, F., M. Winter, D. Bendig, and J. Tuebke. 2023. "Energy Consumption of Current and Future Production of Lithium-Ion and Post Lithium-Ion Battery Cells." *Nature Energy*: 687–691. https://doi.org/10.1038/s41560-023-01355. <Go to ISI>://WOS:001074834300001.

Deng, Hang, Jimin Ye, and Dongmei Huang. 2023. "Design and Analysis of a Galloping Energy Harvester with V-Shape Spring Structure Under Gaussian White Noise." *Chaos Solitons & Fractals* 175. https://doi.org/10.1016/j.chaos.2023.113962. <Go to ISI>://WOS:001080510500001.

Dif, Naas, Elghalia Boudissa, Mhamed Bounekhla, and Ismail Dif. 2020. "Firefly Algorithm Improvement with Application to Induction Machine Parameters Identification." *Revue Roumaine Des Sciences Techniques-Serie Electrotechnique Et Energetique* 65 (1–2): 35–40. <Go to ISI>://WOS:000552052900006.

Dong, Hui. 2020. "Modeling and Simulation of English Speech Rationality Optimization Recognition Based on Improved Particle Filter Algorithm." *Complexity* 2020. https://doi.org/10.1155/2020/6053129. <Go to ISI>://WOS:000570764600003.

Drees, Robin, Frank Lienesch, and Michael Kurrat. 2022. "Durable Fast Charging of Lithium-Ion Batteries Based on Simulations with an Electrode Equivalent Circuit Model." *Batteries-Basel* 8 (4). https://doi.org/10.3390/batteries8040030. <Go to ISI>://WOS:000786232400001.

Du, Banghua, Zhang Yu, Shuhao Yi, Yanlin He, and Yulin Luo. 2021. "State-of-Charge Estimation for Second-Life Lithium-Ion Batteries Based on Cell Difference Model and Adaptive Fading Unscented Kalman Filter Algorithm." *International Journal of Low-Carbon Technologies* 16 (3): 927–939. https://doi.org/10.1093/ijlct/ctab019. <Go to ISI>://WOS:000709550600025.

Du, Changqing, Rui Qi, Zhong Ren, and Di Xiao. 2023. "Research on State-of-Health Estimation for Lithium-Ion Batteries Based on the Charging Phase." *Energies* 16 (3). https://doi.org/10.3390/en16031420. <Go to ISI>://WOS:000930035900001.

Du, Chang-Qing, Jian-Bo Shao, Dong-Mei Wu, Zhong Ren, Zhong-Yi Wu, and Wei-Qun Ren. 2022. "Research on Co-Estimation Algorithm of SOC and

SOH for Lithium-Ion Batteries in Electric Vehicles." *Electronics* 11 (2). https://doi.org/10.3390/electronics11020181. <Go to ISI>://WOS:000747045200001.

Du, Zhekai, Lin Zuo, Jingjing Li, Yu Liu, and Heng Tao Shen. 2022. "Data-Driven Estimation of Remaining Useful Lifetime and State of Charge for Lithium-Ion Battery." *IEEE Transactions on Transportation Electrification* 8 (1): 356–367. https://doi.org/10.1109/tte.2021.3109636. <Go to ISI>://WOS:000792985000031.

Duan, Xiting. 2023. *Parameters Identification and Degradation Diagnosis of Lithium-Ion Battery Using Physical-Based EIS Model*. Edited by Xiting Duan. Ann Arbor: ProQuest Dissertations Publishing.

Duan, Zhijie, Luo Zhang, Lili Feng, Shuguang Yu, Zengyou Jiang, Xiaoming Xu, and Jichao Hong. 2021. "Research on Economic and Operating Characteristics of Hydrogen Fuel Cell Cars Based on Real Vehicle Tests." *Energies* 14 (23). https://doi.org/10.3390/en14237856. <Go to ISI>://WOS:000734872100001.

Fahmy, Youssef A., Weizhong Wang, Alan C. West, and Matthias Preindl. 2021. "Snapshot SoC Identification with Pulse Injection Aided Machine Learning." *Journal of Energy Storage* 41. https://doi.org/10.1016/j.est.2021.102891. <Go to ISI>://WOS:000707774300008.

Falai, Alessandro, Tiziano Alberto Giuliacci, Daniela Anna Misul, and Pier Giuseppe Anselma. 2022. "Reducing the Computational Cost for Artificial Intelligence-Based Battery State-of-Health Estimation in Charging Events." *Batteries-Basel* 8 (11). https://doi.org/10.3390/batteries8110209. <Go to ISI>://WOS:000883865800001.

Fan, Yaxiang, Fei Xiao, Chaoran Li, Guorun Yang, and Xin Tang. 2020. "A Novel Deep Learning Framework for State of Health Estimation of Lithium-Ion Battery." *Journal of Energy Storage* 32. https://doi.org/10.1016/j.est.2020.101741. <Go to ISI>://WOS:000599935900002.

Fan, Yongcun, Haotian Shi, Shunli Wang, Carlos Fernandez, Wen Cao, and Junhan Huang. 2021. "A Novel Adaptive Function-Dual Kalman Filtering Strategy for Online Battery Model Parameters and State of Charge Co-Estimation." *Energies* 14 (8). https://doi.org/10.3390/en14082268. <Go to ISI>://WOS:000644104000001.

Feng, Lei, Lihua Jiang, Jialong Liu, Zhaoyu Wang, Zesen Wei, and Qingsong Wang. 2021. "Dynamic Overcharge Investigations of Lithium Ion Batteries with Different State of Health." *Journal of Power Sources* 507. https://doi.org/10.1016/j.jpowsour.2021.230262. <Go to ISI>://WOS:000685092900006.

Feng, Liang, Jie Ding, and Yiyang Han. 2020. "Improved Sliding Mode Based EKF for the SOC Estimation of Lithium-Ion Batteries." *Ionics* 26 (6): 2875–2882. https://doi.org/10.1007/s11581-019-03368-9. <Go to ISI>://WOS:000516304200001.

Fluegel, Marius, Karsten Richter, Margret Wohlfahrt-Mehrens, and Thomas Waldmann. 2022. "Detection of Li Deposition on Si/Graphite Anodes from Commercial Li-Ion Cells: A Post-Mortem GD-OES Depth Profiling Study." *Journal of the Electrochemical Society* 169 (5). https://doi.org/10.1149/1945-7111/ac70af. <Go to ISI>://WOS:000800663200001.

Fraile Ardanuy, Jesus, Roberto Alvaro-Hermana, Sandra Castano-Solis, and Julia Merino. 2022. "Carbon-Free Electricity Generation in Spain with PV-Storage Hybrid Systems." *Energies* 15 (13). https://doi.org/10.3390/en15134780. <Go to ISI>://WOS:000825630600001.

Fu, Yueshuai, and Huimin Fu. 2023. "A Self-Calibration SOC Estimation Method for Lithium-Ion Battery." *IEEE Access* 11: 37694–37704. https://doi.org/10.1109/access.2023.3266663. <Go to ISI>://WOS:000979584900001.

Fu, Yumeng, Jun Xu, Mingjie Shi, and Xuesong Mei. 2022. "A Fast Impedance Calculation-Based Battery State-of-Health Estimation Method." *IEEE Transactions on Industrial Electronics* 69 (7): 7019–7028. https://doi.org/10.1109/tie.2021.3097668. <Go to ISI>://WOS:000753527500054.

Galos, Joel, Koranat Pattarakunnan, Adam S. Best, Ilias L. Kyratzis, Chun-Hui Wang, and Adrian P. Mouritz. 2021. "Energy Storage Structural Composites with Integrated Lithium-Ion Batteries: A Review." *Advanced Materials Technologies* 6 (8). https://doi.org/10.1002/admt.202001059. <Go to ISI>://WOS:000681425900014.

Ganguli, Souvik, Gagandeep Kaur, and Prasanta Sarkar. 2020. "Identification in the Delta Domain: A Unified Approach Via GWOCFA." *Soft Computing* 24 (7): 4791–4808. https://doi.org/10.1007/s00500-019-04232-8. <Go to ISI>://WOS:000530043200008.

Gao, Yizhao, Kailong Liu, Chong Zhu, Xi Zhang, and Dong Zhang. 2022. "Co-Estimation of State-of-Charge and State-of- Health for Lithium-Ion Batteries Using an Enhanced Electrochemical Model." *IEEE Transactions on Industrial Electronics* 69 (3): 2684–2696. https://doi.org/10.1109/tie.2021.3066946. <Go to ISI>://WOS:000728182300051.

Gao, Yizhao, Chong Zhu, Xi Zhang, and Bangjun Guo. 2021. "Implementation and Evaluation of a Practical Electrochemical-Thermal Model of Lithium-Ion Batteries for EV Battery Management System." *Energy* 221. https://doi.org/10.1016/j.energy.2020.119688. <Go to ISI>://WOS:000632508200008.

Ge, Caian, Yanping Zheng, and Yang Yu. 2022. "State of Charge Estimation of Lithium-Ion Battery Based on Improved Forgetting Factor Recursive Least Squares-Extended Kalman Filter Joint Algorithm." *Journal of Energy Storage* 55. https://doi.org/10.1016/j.est.2022.105474. <Go to ISI>://WOS:000859725000006.

Ge, Yixin, Weixing Gao, Zhengqiu Yuan, Shi Tao, Fanjun Kong, and Bin Qian. 2023. "Synergistic Effect on the Improved Lithium Ion Storage Performance in the Porous Fe2O3@Fe3C@C Composite." *Materials Research Bulletin* 164. https://doi.org/10.1016/j.materresbull.2023.112287. <Go to ISI>://WOS:000985255800001.

Geng, Yuanfei, Hui Pang, and Xiaofei Liu. 2022. "State-of-Charge Estimation for Lithium-Ion Battery Based on PNGV Model and Particle Filter Algorithm." *Journal of Power Electronics* 22 (7): 1154–1164. https://doi.org/10.1007/s43236-022-00422-0. <Go to ISI>://WOS:000778091500003.

Geng, Zeyang, Siyang Wang, Matthew J. Lacey, Daniel Brandell, and Torbjorn Thiringer. 2021. "Bridging Physics-Based and Equivalent Circuit Models for

Lithium-Ion Batteries." *Electrochimica Acta* 372. https://doi.org/10.1016/j.electacta.2021.137829. <Go to ISI>://WOS:000619728100010.

Ghanbari-Adivi, Elham, Mohammad Ehteram, Alireza Farrokhi, and Zohreh Sheikh Khozani. 2022. "Combining Radial Basis Function Neural Network Models and Inclusive Multiple Models for Predicting Suspended Sediment Loads." *Water Resources Management* 36 (11): 4313–4342. https://doi.org/10.1007/s11269-022-03256-4. <Go to ISI>://WOS:000830279900002.

Gholizadeh, Mehdi, and Alireza Yazdizadeh. 2020. "Systematic Mixed Adaptive Observer and EKF Approach to Estimate SOC and SOH of Lithium-Ion Battery." *IET Electrical Systems in Transportation* 10 (2): 135–143. https://doi.org/10.1049/iet-est.2019.0033. <Go to ISI>://WOS:000535922200002.

Gong, Qingrui, Ping Wang, and Ze Cheng. 2022. "A Data-Driven Model Framework Based on Deep Learning for Estimating the States of Lithium-Ion Batteries." *Journal of the Electrochemical Society* 169 (3). https://doi.org/10.1149/1945-7111/ac5bac. <Go to ISI>://WOS:000770132500001.

Gong, Yadong, Xiaoyong Zhang, Dianzhu Gao, Heng Li, Lisen Yan, Jun Peng, and Zhiwu Huang. 2022. "State-of-Health Estimation of Lithium-Ion Batteries Based on Improved Long Short-Term Memory Algorithm." *Journal of Energy Storage* 53. https://doi.org/10.1016/j.est.2022.105046. <Go to ISI>://WOS:000814764200002.

Gu, Tianyu, Jie Sheng, Qiuhua Fan, and Dongqing Wang. 2022. "The Modified Multi-Innovation Adaptive EKF Algorithm for Identifying Battery SOC." *Ionics* 28 (8): 3877–3891. https://doi.org/10.1007/s11581-022-04603-6. <Go to ISI>://WOS:000801216700001.

Guan, Peiyuan, Jie Min, Fandi Chen, Shuo Zhang, Long Hu, Zhipeng Ma, Zhaojun Han, Lu Zhou, Haowei Jia, Yunjian Liu, Neeraj Sharma, Dawei Su, Judy N. Hart, Tao Wan, and Dewei Chu. 2023. "Enhancing the Electrochemical Properties of Nickel-Rich Cathode by Surface Coating with Defect-Rich Strontium Titanate." *ACS Applied Materials & Interfaces* 15 (24): 29308–29320. https://doi.org/10.1021/acsami.3c04344. <Go to ISI>://WOS:001010274300001.

Gudise, Sandhya, K. Giri Babu, and T. Satya Savithri. 2023. "An Advanced Fuzzy C-Means Algorithm for the Tissue Segmentation from Brain Magnetic Resonance Images in the Presence of Noise and Intensity Inhomogeneity." *Imaging Science Journal*. https://doi.org/10.1080/13682199.2023.2210400. <Go to ISI>://WOS:000990176300001.

Gumuscu, Abdulkadir, Serkan Kaya, Mehmet Emin Tenekeci, Izzettin Hakan Karacizmeli, and Ibrahim Berkan Aydilek. 2022. "The Impact of Local Search Strategies on Chaotic Hybrid Firefly Particle Swarm Optimization Algorithm in Flow-Shop Scheduling." *Journal of King Saud University-Computer and Information Sciences* 34 (8): 6432–6440. https://doi.org/10.1016/j.jksuci.2021.07.017. <Go to ISI>://WOS:000862928800019.

Gun'ko, Yu L., O. L. Kozina, A. A. Myunts, N. O. Kuzyakin, E. Yu Ananieva, and M. G. Mikhalenko. 2020. "Digital Simulation of Discharge of Nickel-Cadmium Batteries." *Russian Journal of Electrochemistry* 56 (12): 997–1010. https://doi.org/10.1134/s1023193520120071. <Go to ISI>://WOS:000613481000006.

Guo, Qing, Youqian Liu, and Luxin Cai. 2023. "An Experimental Study on the Potential Purchase Behavior of Chinese Consumers of New Energy Hybrid Electric Vehicles." *Frontiers in Environmental Science* 11. https://doi.org/10.3389/fenvs.2023.1159846. <Go to ISI>://WOS:000973079400001.

Guo, Shanshan, Zhiqiang Han, Jun Wei, Shenggang Guo, and Liang Ma. 2022. "A Novel DC-AC Fast Charging Technology for Lithium-Ion Power Battery at Low-Temperatures." *Sustainability* 14 (11). https://doi.org/10.3390/su14116544. <Go to ISI>://WOS:000809001700001.

Guo, Shanshan, and Liang Ma. 2023. "A Comparative Study of Different Deep Learning Algorithms for Lithium-Ion Batteries on State-of-Charge Estimation." *Energy* 263. https://doi.org/10.1016/j.energy.2022.125872. <Go to ISI>://WOS:000879221900006.

Gutsch, Moritz, and Jens Leker. 2022. "Review Article Global Warming Potential of Lithium-Ion Battery Energy Storage Systems: A Review." *Journal of Energy Storage* 52. https://doi.org/10.1016/j.est.2022.105030. <Go to ISI>://WOS:000815975200001.

Hammou, Abdelilah, Raffaele Petrone, Demba Diallo, and Hamid Gualous. 2023. "Estimating the Health Status of Li-Ion NMC Batteries From Energy Characteristics for EV Applications." *IEEE Transactions on Energy Conversion* 38 (3): 2160–2168. https://doi.org/10.1109/tec.2023.3259744. <Go to ISI>://WOS:001081209800058.

Han, Junyan, Xiaoyuan Wang, Huili Shi, Bin Wang, Gang Wang, Longfei Chen, and Quanzheng Wang. 2022. "Research on the Impacts of Vehicle Type on Car-Following Behavior, Fuel Consumption and Exhaust Emission in the V2X Environment." *Sustainability* 14 (22). https://doi.org/10.3390/su142215231. <Go to ISI>://WOS:000887616400001.

Han, Youjun, Hongyuan Yuan, Ying Shao, Jin Li, and Xuejie Huang. 2023. "Capacity Consistency Prediction and Process Parameter Optimization of Lithium-Ion Battery Based on Neural Network and Particle Swarm Optimization Algorithm." *Advanced Theory and Simulations*. https://doi.org/10.1002/adts.202300125. <Go to ISI>://WOS:000991789400001.

Hao, Xueyi, Shunli Wang, Yongcun Fan, Yanxin Xie, and Carlos Fernandez. 2023. "An Improved Forgetting Factor Recursive Least Square and Unscented Particle Filtering Algorithm for Accurate Lithium-Ion Battery State of Charge Estimation." *Journal of Energy Storage* 59. https://doi.org/10.1016/j.est.2022.106478. <Go to ISI>://WOS:000930750900001.

Hassan, Bryar A. 2021. "CSCF: A Chaotic Sine Cosine Firefly Algorithm for Practical Application Problems." *Neural Computing & Applications* 33 (12): 7011–7030. https://doi.org/10.1007/s00521-020-05474-6. <Go to ISI>://WOS:000591218300001.

Hassan, Mohamed H., Salah Kamel, Ahmad Eid, Loai Nasrat, Francisco Jurado, and Mohamed F. Elnaggar. 2023. "A Developed Eagle-Strategy Supply-Demand Optimizer for Solving Economic Load Dispatch Problems." *Ain Shams Engineering Journal* 14 (5). https://doi.org/10.1016/j.asej.2022.102083. <Go to ISI>://WOS:000915307500001.

He, Mingfang, Shunli Wang, Carlos Fernandez, Chunmei Vu, Xiaoxia Li, and Etse Dablu Bobobee. 2021. "A Novel Adaptive Particle Swarm Optimization Algorithm Based High Precision Parameter Identification and State Estimation of Lithium-Ion Battery." *International Journal of Electrochemical Science* 16 (5). https://doi.org/10.20964/2021.05.55. <Go to ISI>://WOS:000661483700004.

Heimes, Heiner Hans, Christian Offermanns, Ahmad Mohsseni, Hendrik Laufen, Uwe Westerhoff, Louisa Hoffmann, Philip Niehoff, Michael Kurrat, Martin Winter, and Achim Kampker. 2020. "The Effects of Mechanical and Thermal Loads During Lithium-Ion Pouch Cell Formation and Their Impacts on Process Time." *Energy Technology* 8 (2). https://doi.org/10.1002/ente.201900118. <Go to ISI>://WOS:000511910200023.

Heins, Tom Patrick, Nicolas Schlueter, Sabine Teresa Ernst, and Uwe Schroeder. 2020. "On the Interpretation of Impedance Spectra of Large-Format Lithium-Ion Batteries and Its Application in Aging Studies." *Energy Technology* 8 (2). https://doi.org/10.1002/ente.201900279. <Go to ISI>://WOS:000511910200020.

Hong, Jianwang, Ricardo A. Ramirez-Mendoza, and Jorge de J. Lozoya-Santos. 2020. "Adjustable Scaling Parameters for State of Charge Estimation for Lithium-Ion Batteries Using Iterative Multiple UKFs." *Mathematical Problems in Engineering* 2020. https://doi.org/10.1155/2020/4037306. <Go to ISI>://WOS:000530332000008.

Hong, Sheng, Tianyu Yue, and Hao Liu. 2022. "Vehicle Energy System Active Defense: A Health Assessment of Lithium-Ion Batteries." *International Journal of Intelligent Systems* 37 (12): 10081–10099. https://doi.org/10.1002/int.22309. <Go to ISI>://WOS:000577382900001.

Honrao, Shreyas J., Xin Yang, Balachandran Radhakrishnan, Shigemasa Kuwata, Hideyuki Komatsu, Atsushi Ohma, Maarten Sierhuis, and John W. Lawson. 2021. "Discovery of Novel Li SSE and Anode Coatings Using Interpretable Machine Learning and High-Throughput Multi-Property Screening." *Scientific Reports* 11 (1). https://doi.org/10.1038/s41598-021-94275-5. <Go to ISI>://WOS:000686717900008.

Hu, Jingwei, Bing Lin, Mingfen Wang, Jie Zhang, Wenliang Zhang, and Yu Lu. 2022. "State of Charge Centralized Estimation of Road Condition Information Based on Fuzzy Sunday Algorithm." *Energies* 15 (8). https://doi.org/10.3390/en15082853. <Go to ISI>://WOS:000785283200001.

Hu, Liwen, Wenbo Wang, and Guorong Ding. 2023. "RUL Prediction for Lithium-Ion Batteries Based on Variational Mode Decomposition and Hybrid Network Model." *Signal Image and Video Processing* 17 (6): 3109–3117. https://doi.org/10.1007/s11760-023-02532-z. <Go to ISI>://WOS:000982757000001.

Hu, Minghui, Yunxiao Li, Shuxian Li, Chunyun Fu, Datong Qin, and Zonghua Li. 2018. "Lithium-Ion Battery Modeling and Parameter Identification Based on Fractional Theory." *Energy* 165: 153–163. https://doi.org/10.1016/j.energy.2018.09.101. <Go to ISI>://WOS:000455171600013.

Hu, Panpan, W. F. Tang, C. H. Li, Shu-Lun Mak, C. Y. Li, and C. C. Lee. 2023. "Joint State of Charge (SOC) and State of Health (SOH) Estimation for Lithium-Ion Batteries Packs of Electric Vehicles Based on NSSR-LSTM Neural Network." *Energies* 16 (14). https://doi.org/10.3390/en16145313. <Go to ISI>://WOS:001038666000001.

Hu, Wangyang, and Shaishai Zhao. 2022. "Remaining Useful Life Prediction of Lithium-Ion Batteries Based on Wavelet Denoising and Transformer Neural Network." *Frontiers in Energy Research* 10. https://doi.org/10.3389/fenrg.2022.969168. <Go to ISI>://WOS:000845011400001.

Hu, Xiaosong, Yunhong Che, Xianke Lin, and Zhongwei Deng. 2020. "Health Prognosis for Electric Vehicle Battery Packs: A Data-Driven Approach." *IEEE-ASME Transactions on Mechatronics* 25 (6): 2622–2632. https://doi.org/10.1109/tmech.2020.2986364. <Go to ISI>://WOS:000599503600004.

Hu, Zhihua, Qingke Huang, Wenqin Cai, Zeng Zeng, Kai Chen, Yan Sun, Qingquan Kong, Wei Feng, Ke Wang, Zhenguo Wu, Yang Song, and Xiaodong Guo. 2023. "Research Progress on Enhancing the Performance of High Nickel Single Crystal Cathode Materials for Lithium-Ion Batteries." *Industrial & Engineering Chemistry Research*. https://doi.org/10.1021/acs.iecr.2c04021. <Go to ISI>://WOS:000930568200001.

Huang, Cong-Sheng. 2023. "An Online Condition-Based Parameter Identification Switching Algorithm for Lithium-Ion Batteries in Electric Vehicles." *IEEE Transactions on Vehicular Technology* 72 (2): 1701–1709. https://doi.org/10.1109/tvt.2022.3210688. <Go to ISI>://WOS:000944202400025.

Huang, Guoyong, Xi Dong, Jianwei Du, Xiaohua Sun, Botian Li, and Haimu Ye. 2021. "High-Voltage Electrolyte for Lithium-Ion Batteries." *Progress in Chemistry* 33 (5): 855–867. https://doi.org/10.7536/pc200634. <Go to ISI>://WOS:000664800300010.

Huang, K. David, Zhong-Ting Cao, Jyun-Ming Jhang, Cheng-Jung Yang, and Po-Tuan Chen. 2022. "Parameter Improvement of Composite Sinusoidal Waveform Charging Strategy for Reviving Lithium-Ion Batteries Capacity." *Journal of the Chinese Society of Mechanical Engineers* 43 (3): 209–216. <Go to ISI>://WOS:000840826800002.

Huang, Kui, Hao Xiong, Haili Dong, Yuling Liu, Yuanhuan Lu, Kunjie Liu, and Junzhen Wang. 2022. "Carbon Thermal Reduction of Waste Ternary Cathode Materials and Wet Magnetic Separation Based on Ni/MnO Nanocomposite Particles." *Process Safety and Environmental Protection* 165: 278–285. https://doi.org/10.1016/j.psep.2022.07.012. <Go to ISI>://WOS:000830885000005.

Huang, Rui, Yidan Xu, Qichao Wu, Junxuan Chen, Fenfang Chen, and Xiaoli Yu. 2023. "Simulation Study on Heat Generation Characteristics of Lithium-Ion Battery Aging Process." *Electronics* 12 (6). https://doi.org/10.3390/electronics12061444. <Go to ISI>://WOS:000955766100001.

Huo, Weiwei, Yunxu Jia, Yong Chen, and Aobo Wang. 2023. "Joint Estimation for SOC and Capacity After Current Measurement Offset Redress with Two-Stage Forgetting Factor Recursive Least Square Method." *Journal of Power*

*Electronics*. https://doi.org/10.1007/s43236-023-00683-3. <Go to ISI>://WOS:001046397500003.

Ibrahim, Tarek, Tamas Kerekes, Dezso Sera, Shahrzad S. Mohammadshahi, and Daniel-Ioan Stroe. 2023. "Sizing of Hybrid Supercapacitors and Lithium-Ion Batteries for Green Hydrogen Production from PV in the Australian Climate." *Energies* 16 (5). https://doi.org/10.3390/en16052122. <Go to ISI>://WOS:000947246900001.

Iurilli, Pietro, Claudio Brivio, Rafael E. Carrillo, and Vanessa Wood. 2022. "Physics-Based SoH Estimation for Li-Ion Cells." *Batteries-Basel* 8 (11). https://doi.org/10.3390/batteries8110204. <Go to ISI>://WOS:000883139300001.

Jafari, Sadiqa, and Yung-Cheol Byun. 2022. "XGBoost-Based Remaining Useful Life Estimation Model with Extended Kalman Particle Filter for Lithium-Ion Batteries." *Sensors* 22 (23). https://doi.org/10.3390/s22239522. <Go to ISI>://WOS:000897419300001.

Janga, Vijaykumar, and Srinivasa Reddy Edara. 2021. "Epilepsy and Seizure Detection Using JLTM Based ICFFA and Multiclass SVM Classifier." *Traitement Du Signal* 38 (3): 883–893. https://doi.org/10.18280/ts.380335. <Go to ISI>://WOS:000681761900035.

Jerouschek, Daniel, Oemer Tan, Ralph Kennel, and Ahmet Taskiran. 2020. "Data Preparation and Training Methodology for Modeling Lithium-Ion Batteries Using a Long Short-Term Memory Neural Network for Mild-Hybrid Vehicle Applications." *Applied Sciences-Basel* 10 (21). https://doi.org/10.3390/app10217880. <Go to ISI>://WOS:000589009300001.

Ji, Hao, Wei Zhang, Xu-Hai Pan, Min Hua, Yi-Hong Chung, Chi-Min Shu, and Li-Jing Zhang. 2020. "State of Health Prediction Model Based on Internal Resistance." *International Journal of Energy Research* 44 (8): 6502–6510. https://doi.org/10.1002/er.5383. <Go to ISI>://WOS:000525822300001.

Ji, Weikang, Shunli Wang, Chuanyun Zou, and Haotian Shi. 2021. "A Novel Fading Memory Square Root UKF Algorithm for the High-precision State of Charge Estimation of High-Power Lithium-Ion Batteries." *International Journal of Electrochemical Science* 16 (7). https://doi.org/10.20964/2021.07.68. <Go to ISI>://WOS:000669352100037.

Jia, Xianyi, Shunli Wang, Jialu Qiao, and Wen Cao. 2022. "An Adaptive Spherical Square-Root Double Unscented Kalman Filtering Algorithm for Estimating State-of-Charge of Lithium-Ion Batteries." *International Journal of Energy Research* 46 (10): 14256–14267. https://doi.org/10.1002/er.8139. <Go to ISI>://WOS:000799492700001.

Jia, Xinyu, Caiping Zhang, Leyi Wang, Linjing Zhang, and Xinzhen Zhou. 2022. "Early Diagnosis of Accelerated Aging for Lithium-Ion Batteries With an Integrated Framework of Aging Mechanisms and Data-Driven Methods." *IEEE Transactions on Transportation Electrification* 8 (4): 4722–4742. https://doi.org/10.1109/tte.2022.3180805. <Go to ISI>://WOS:000871082600057.

Jia, Yikai, Xiang Gao, Lin Ma, and Jun Xu. 2023. "Comprehensive Battery Safety Risk Evaluation: Aged Cells Versus Fresh Cells Upon Mechanical Abusive Loadings." *Advanced Energy Materials* 13 (24). https://doi.org/10.1002/aenm.202300368. <Go to ISI>://WOS:000984139100001.

Jiang, Bo, Jiangong Zhu, Xueyuan Wang, Xuezhe Wei, Wenlong Shang, and Haifeng Dai. 2022. "A Comparative Study of Different Features Extracted from Electrochemical Impedance Spectroscopy in State of Health Estimation for Lithium-Ion Batteries." *Applied Energy* 322. https://doi.org/10.1016/j.apenergy.2022.119502. <Go to ISI>://WOS:000833364400005.

Jiang, Cong, Shunli Wang, Bin Wu, Bobobee Etse-Dabu, and Xin Xiong. 2020. "A Novel Adaptive Extended Kalman Filtering and Electrochemical-Circuit Combined Modeling Method for the Online Ternary Battery State-of-Charge Estimation." *International Journal of Electrochemical Science* 15 (10): 9720–9733. https://doi.org/10.20964/2020.10.09. <Go to ISI>://WOS:000580810400014.

Jiang, Lixue, Chunwei Dong, Bo Jin, Zi Wen, and Qing Jiang. 2019. "$ZnFe_2O_4$@PPy Core-Shell Structure for High-Rate Lithium-Ion Storage." *Journal of Electroanalytical Chemistry* 851. https://doi.org/10.1016/j.jelechem.2019.113442. <Go to ISI>://WOS:000496838700012.

Jiang, Nan, and Hui Pang. 2022. "Study on Co-Estimation of SoC and SoH for Second-Use Lithium-Ion Power Batteries." *Electronics* 11 (11). https://doi.org/10.3390/electronics11111789. <Go to ISI>://WOS:000809130600001.

Jiang, Yinfeng, and Wenxiang Song. 2023. "Predicting the Cycle Life of Lithium-Ion Batteries Using Data-Driven Machine Learning Based on Discharge Voltage Curves." *Batteries-Basel* 9 (8). https://doi.org/10.3390/batteries9080413. <Go to ISI>://WOS:001057681500001.

Jiang, Yuanyuan, Jie Zhang, Ling Xia, and Yanbin Liu. 2020. "State of Health Estimation for Lithium-Ion Battery Using Empirical Degradation and Error Compensation Models." *IEEE Access* 8: 123858–123868. https://doi.org/10.1109/access.2020.3005229. <Go to ISI>://WOS:000554575000001.

Jiao, Shiqin, Guiyang Zhang, Mei Zhou, and Guoqi Li. 2023. "A Comprehensive Review of Research Hotspots on Battery Management Systems for UAVs." *IEEE Access* 11: 84636–84650. https://doi.org/10.1109/access.2023.3301989. <Go to ISI>://WOS:001049938100001.

Jiaqiang, E., Bin Zhang, Yan Zeng, Ming Wen, Kexiang Wei, Zhonghua Huang, Jingwei Chen, Hao Zhu, and Yuanwang Deng. 2022. "Effects Analysis on Active Equalization Control of Lithium-Ion Batteries Based on Intelligent Estimation of the State-of-Charge." *Energy* 238. https://doi.org/10.1016/j.energy.2021.121822. <Go to ISI>://WOS:000701782300001.

Jo, Sungwoo, Sunkyu Jung, and Taemoon Roh. 2021. "Battery State-of-Health Estimation Using Machine Learning and Preprocessing with Relative State-of-Charge." *Energies* 14 (21). https://doi.org/10.3390/en14217206. <Go to ISI>://WOS:000726544300001.

Kacica, Clayton T. 2020. *Scalable Synthesis of Nanostructured Metal Oxide Films Using Aerosol Chemical Vapor Deposition for Energy Storage Applications*. Edited by Clayton T Kacica. Ann Arbor: ProQuest Dissertations Publishing.

Kahnamouei, Ali Shakeri, and Saeed Lotfifard. 2023. "Optimized Fault Location Identification in Power Distribution Systems With Inverter-Interfaced

Distributed Generations." *IEEE Transactions on Power Delivery* 38 (5): 3429–3440. https://doi.org/10.1109/tpwrd.2023.3286861. <Go to ISI>://WOS:001075302900040.

Kang, Jiayi, Xiuqiong Chen, Yangtianze Tao, and Stephen Shing-Toung Yau. 2022. "Optimal Transportation Particle Filter for Linear Filtering Systems With Correlated Noises." *IEEE Transactions on Aerospace and Electronic Systems* 58 (6): 5190–5203. https://doi.org/10.1109/taes.2022.3166863. <Go to ISI>://WOS:000895081000025.

Karimi, Danial, Hamidreza Behi, Joeri Van Mierlo, and Maitane Berecibar. 2022. "A Comprehensive Review of Lithium-Ion Capacitor Technology: Theory, Development, Modeling, Thermal Management Systems, and Applications." *Molecules* 27 (10). https://doi.org/10.3390/molecules27103119. <Go to ISI>://WOS:000803286200001.

Kawahara, Yohei, Kei Sakabe, Ryohei Nakao, Kenichiro Tsuru, Keiichiro Okawa, Yoshinori Aoshima, Akihiko Kudo, and Akihiko Emori. 2021. "Development of Status Detection Method of Lithium-Ion Rechargeable Battery for Hybrid Electric Vehicles." *Journal of Power Sources* 481. https://doi.org/10.1016/j.jpowsour.2020.228760. <Go to ISI>://WOS:000593860600003.

Kaya, Serkan, Abdulkadir Gumuscu, Ibrahim Berkan Aydilek, Izzettin Hakan Karacizmeli, and Mehmet Emin Tenekeci. 2021. "Solution for Flow Shop Scheduling Problems Using Chaotic Hybrid Firefly and Particle Swarm Optimization Algorithm with Improved local Search." *Soft Computing* 25 (10): 7143–7154. https://doi.org/10.1007/s00500-021-05673-w. <Go to ISI>://WOS:000627238500001.

Kheirkhah-Rad, Ehsan, Amirreza Parvareh, Moein Moeini-Aghtaie, and Payman Dehghanian. 2023. "A Data-Driven State-of-Health Estimation Model for Lithium-Ion Batteries Using Referenced-Based Charging Time." *IEEE Transactions on Power Delivery* 38 (5): 3406–3416. https://doi.org/10.1109/tpwrd.2023.3276268. <Go to ISI>://WOS:001075302900038.

Kim, Jonghyeon, and Julia Kowal. 2021. "A Method for Monitoring State-of-Charge of Lithium-Ion Cells Using Multi-Sine Signal Excitation." *Batteries-Basel* 7 (4). https://doi.org/10.3390/batteries7040076. <Go to ISI>://WOS:000735847800001.

Kim, Jonghyeon, and Julia Kowal. 2022. "Development of a Matlab/Simulink Model for Monitoring Cell State-of-Health and State-of-Charge Via Impedance of Lithium-Ion Battery Cells." *Batteries-Basel* 8 (2). https://doi.org/10.3390/batteries8020008. <Go to ISI>://WOS:000767609400001.

Kim, Min Young, Young-Woong Song, Jinsub Lim, Sang Jun Park, Byeong-Su Kang, Youngsun Hong, Ho Sung Kim, and Jong Hun Han. 2022. "LATP-Coated NCM-811 for High-Temperature Operation of All-Solid Lithium Battery." *Materials Chemistry and Physics* 290. https://doi.org/10.1016/j.matchemphys.2022.126644. <Go to ISI>://WOS:000844865900003.

Kim, Minho, and Soohee Han. 2021. "Novel Data-Efficient Mechanism-Agnostic Capacity Fade Model for Li-Ion Batteries." *IEEE Transactions on Industrial Electronics* 68 (7): 6267–6275. https://doi.org/10.1109/tie.2020.2996156. <Go to ISI>://WOS:000633442800073.

Kim, Si-Jin, L. E. E. Jong-Hyun, Dong-Hun Wang, and In Soo Lee. 2021. "Vehicle Simulator and SOC Estimation of Battery using Artificial Neural Networks." *Journal of Korean Institute of Information Technology* 19 (5): 51–62. https://doi.org/10.14801/jkiit.2021.19.5.51. <Go to ISI>://KJD:ART002720504.

Kong, Depeng, Shuhui Wang, and Ping Ping. 2021. "A Novel Parameter Adaptive Method for State of Charge Estimation of Aged Lithium Batteries." *Journal of Energy Storage* 44. https://doi.org/10.1016/j.est.2021.103389. <Go to ISI>://WOS:000711640300003.

Kumar, Kartik, and Kapil Pareek. 2023. "Fast Charging of Lithium-Ion Battery Using Multistage Charging and Optimization with Grey Relational Analysis." *Journal of Energy Storage* 68. https://doi.org/10.1016/j.est.2023.107704. <Go to ISI>://WOS:000999159800001.

Kumari, Pooja, Ashutosh Kumar Singh, and Niranjan Kumar. 2023. "Optimized Deep Learning Strategy for Estimation of State of Charge at Different C-Rate with Varying Temperature." *Electrical Engineering*. https://doi.org/10.1007/s00202-023-01925-0. <Go to ISI>://WOS:001027453500001.

Kurzweil, Peter, Bernhard Frenzel, and Wolfgang Scheuerpflug. 2022. "A Novel Evaluation Criterion for the Rapid Estimation of the Overcharge and Deep Discharge of Lithium-Ion Batteries Using Differential Capacity." *Batteries-Basel* 8 (8). https://doi.org/10.3390/batteries8080086. <Go to ISI>://WOS:000846120300001.

Kurzweil, Peter, Wolfgang Scheuerpflug, Bernhard Frenzel, Christian Schell, and Josef Schottenbauer. 2022. "Differential Capacity as a Tool for SOC and SOH Estimation of Lithium Ion Batteries Using Charge/Discharge Curves, Cyclic Voltammetry, Impedance Spectroscopy, and Heat Events: A Tutorial." *Energies* 15 (13). https://doi.org/10.3390/en15134520. <Go to ISI>://WOS:000822289600001.

Lai, Xin, Ming Yuan, Xiaopeng Tang, Yi Yao, Jiahui Weng, Furong Gao, Weiguo Ma, and Yuejiu Zheng. 2022. "Co-Estimation of State-of-Charge and State-of-Health for Lithium-Ion Batteries Considering Temperature and Ageing." *Energies* 15 (19). https://doi.org/10.3390/en15197416. <Go to ISI>://WOS:000866763000001.

Lange, Theresa. 2022. "Derivation of Ensemble Kalman-Bucy Filters with Unbounded Nonlinear Coefficients." *Nonlinearity* 35 (2): 1061–1092. https://doi.org/10.1088/1361-6544/ac4337. <Go to ISI>://WOS:000739640800001.

Lee, Gyumin, Juram Kim, and Changyong Lee. 2022. "State-of-Health Estimation of Li-Ion Batteries in the Early Phases of Qualification Tests: An Interpretable Machine Learning Approach." *Expert Systems with Applications* 197. https://doi.org/10.1016/j.eswa.2022.116817. <Go to ISI>://WOS:000819798800010.

Lee, Jaewoo, Dongcheul Lee, Chee Burm Shin, So-Yeon Lee, Seung Mi Oh, Jung-Je Woo, and Jang Ilchan. 2021. "Modeling to Estimate the Cycle Life of a Lithium-ion Battery." *Korean Chemical Engineering Research* 59 (3): 393–398. <Go to ISI>://KJD:ART002741384.

Lee, Kwang-Jae, Won-Hyung Lee, and Kwang-Ki K. Kim. 2023. "Battery State-of-Charge Estimation Using Data-Driven Gaussian Process Kalman Filters."

*Journal of Energy Storage* 72. https://doi.org/10.1016/j.est.2023.108392. <Go to ISI>://WOS:001059429900001.

Lee, Suhak. 2021. *Electrode-Specific Degradation Diagnostics for Lithium-Ion Batteries with Practical Considerations.* Edited by Suhak Lee. Ann Arbor: ProQuest Dissertations Publishing.

Leon, Evan M., and Shelie A. Miller. 2020. "An Applied Analysis of the Recyclability of Electric Vehicle Battery Packs." *Resources Conservation and Recycling* 157. https://doi.org/10.1016/j.resconrec.2020.104593. <Go to ISI>://WOS:000540606400005.

Levieux-Souid, Yanis, Jean-Frederic Martin, Philippe Moreau, Nathalie Herlin-Boime, and Sophie Le Caer. 2022. "Radiolysis of Electrolytes in Batteries: A Quick and Efficient Screening Process for the Selection of Electrolyte-Additive Formulations." *Small Methods* 6 (10). https://doi.org/10.1002/smtd.202200712. <Go to ISI>://WOS:000844222500001.

Li, Bing. 2020. *Examination of Lithium-Ion Battery Performance Degradation Under Dynamic Environment and Early Detection of Thermal Runaway with Internal Sensor Measurement.* Edited by Bing Li. Ann Arbor: ProQuest Dissertations Publishing.

Li, Bohao, and Chunsheng Hu. 2022. "Multifunctional Estimation and Analysis of Lithium-Ion Battery State Based on Data Model Fusion under Multiple Constraints." *Journal of the Electrochemical Society* 169 (11). https://doi.org/10.1149/1945-7111/aca2ee. <Go to ISI>://WOS:000893309000001.

Li, Dezhi, Shuo Li, Shubo Zhang, Jianrui Sun, Licheng Wang, and Kai Wang. 2022. "Aging State Prediction for Supercapacitors Based on Heuristic Kalman Filter Optimization Extreme Learning Machine." *Energy* 250. https://doi.org/10.1016/j.energy.2022.123773. <Go to ISI>://WOS:000792615600007.

Li, Hao, Weige Zhang, Bingxiang Sun, Jiuchun Jiang, Ze Yin, Jian Wu, and Xitian He. 2022. "Lithium-Ion Battery Modeling Under High-Frequency Ripple Current for Co-Simulation of High-Power DC-DC Converters." *Journal of Energy Storage* 54. https://doi.org/10.1016/j.est.2022.105284. <Go to ISI>://WOS:000893161400004.

Li, Huan, Shunli Wang, Monirul Islam, Etse Dablu Bobobee, Chuanyun Zou, and Carlos Fernandez. 2022. "A Novel State of Charge Estimation Method of Lithium-Ion Batteries Based on the IWOA-AdaBoost-Elman Algorithm." *International Journal of Energy Research* 46 (4): 5134–5151. https://doi.org/10.1002/er.7505. <Go to ISI>://WOS:000729130500001.

Li, Huan, Chuanyun Zou, Carlos Fernandez, Shunli Wang, Yongcun Fan, and Donglei Liu. 2021. "A Novel State of Charge Estimation for Energy Storage Systems Based on the Joint NARX Network and Filter Algorithm." *International Journal of Electrochemical Science* 16 (12). https://doi.org/10.20964/2021.12.50. <Go to ISI>://WOS:000756422800005.

Li, Huanhuan, Zhiwei Qu, Tong Xu, Yaping Wang, Xiaosong Fan, Haobin Jiang, Chaochun Yuan, and Long Chen. 2022. "SOC Estimation Based on the Gas-Liquid Dynamics Model Using Particle Filter Algorithm." *International Journal of Energy Research* 46 (15): 22913–22925. https://doi.org/10.1002/er.8594. <Go to ISI>://WOS:000858375500001.

Li, Jiabo, Min Ye, Xiaokang Ma, Qiao Wang, and Yan Wang. 2023. "SOC Estimation and Fault Diagnosis Framework of Battery Based on Multi-Model Fusion Modeling." *Journal of Energy Storage* 65. https://doi.org/10.1016/j.est.2023.107296. <Go to ISI>://WOS:000984083100001.

Li, Jia-Jing, Yang Dai, and Jin-Cheng Zheng. 2021. "Strain Engineering of Ion Migration in $LiCoO_2$." *Frontiers of Physics* 17 (1). https://doi.org/10.1007/s11467-021-1086-5. <Go to ISI>://WOS:000692180600001.

Li, Jiang, Lihong Guo, Yan Li, Chang Liu, Lijuan Wang, and Hui Hu. 2021. "Enhancing Whale Optimization Algorithm with Chaotic Theory for Permutation Flow Shop Scheduling Problem." *International Journal of Computational Intelligence Systems* 14 (1): 651–675. https://doi.org/10.2991/ijcis.d.210112.002. <Go to ISI>://WOS:000619260100028.

Li, Jiarui, Xiaofan Huang, Xiaoping Tang, Jinhua Guo, Qiying Shen, Yuan Chai, Wu Lu, Tong Wang, and Yongsheng Liu. 2023. "The State-of-Charge Predication of Lithium-Ion Battery Energy Storage System Using Data-Driven Machine Learning." *Sustainable Energy Grids & Networks* 34. https://doi.org/10.1016/j.segan.2023.101020. <Go to ISI>://WOS:001010774700001.

Li, Junfu, Shaochun Xu, Changsong Dai, Ming Zhao, and Zhenbo Wang. 2022. "Characteristic Prediction and Temperature-Control Strategy Under Constant Power Conditions for Lithium-Ion Batteries." *Batteries-Basel* 8 (11). https://doi.org/10.3390/batteries8110217. <Go to ISI>://WOS:000883420800001.

Li, Ling-Ling, Zhi-Feng Liu, and Ching-Hsin Wang. 2020. "The Open-Circuit Voltage Characteristic and State of Charge Estimation for Lithium-Ion Batteries Based on an Improved Estimation Algorithm." *Journal of Testing and Evaluation* 48 (2): 1712–1730. https://doi.org/10.1520/jte20170558. <Go to ISI>://WOS:000527609100056.

Li, Ning, Fuxing He, Wentao Ma, Ruotong Wang, Lin Jiang, and Xiaoping Zhang. 2022. "An Indirect State-of-Health Estimation Method Based on Improved Genetic and Back Propagation for Online Lithium-Ion Battery Used in Electric Vehicles." *IEEE Transactions on Vehicular Technology* 71 (12): 12682–12690. https://doi.org/10.1109/tvt.2022.3196225. <Go to ISI>://WOS:000908826000021.

Li, Ran, Xue Wei, Hui Sun, Hao Sun, and Xiaoyu Zhang. 2022. "Fast Charging Optimization for Lithium-Ion Batteries Based on Improved Electro-Thermal Coupling Model." *Energies* 15 (19). https://doi.org/10.3390/en15197038. <Go to ISI>://WOS:000867986300001.

Li, Renzheng, Jichao Hong, Huaqin Zhang, and Xinbo Chen. 2022. "Data-Driven Battery State of Health Estimation Based on Interval Capacity for Real-World Electric Vehicles." *Energy* 257. https://doi.org/10.1016/j.energy.2022.124771. <Go to ISI>://WOS:000853695300001.

Li, Shuangqi, Hongwen He, and Jianwei Li. 2019. "Big Data Driven Lithium-Ion Battery Modeling Method Based on SDAE-ELM Algorithm and Data Pre-Processing Technology." *Applied Energy* 242: 1259–1273. https://doi.org/10.1016/j.apenergy.2019.03.154. <Go to ISI>://WOS:000470045800095.

Li, Wenhua, Zhipeng Jiao, Qian Xiao, Jinhao Meng, Yunfei Mu, Hongjie Jia, Remus Teodorescu, and Frede Blaabjerg. 2019. "A Study on Performance Characterization Considering Six-Degree-of-Freedom Vibration Stress and Aging Stress for Electric Vehicle Battery Under Driving Conditions." *IEEE Access* 7: 112180–112190. https://doi.org/10.1109/access.2019.2935380. <Go to ISI>://WOS:000482831700002.

Li, Wenqian, Yan Yang, Dongqing Wang, and Shengqiang Yin. 2020. "The Multi-Innovation Extended Kalman Filter Algorithm for Battery SOC Estimation." *Ionics* 26 (12): 6145–6156. https://doi.org/10.1007/s11581-020-03716-0. <Go to ISI>://WOS:000561515200001.

Li, Xiao, Ai Jie Cheng, and Hai Xiang Lin. 2021. "Sample Regenerating Particle Filter Combined With Unequal Weight Ensemble Kalman Filter for Nonlinear Systems." *IEEE Access* 9: 109612–109623. https://doi.org/10.1109/access.2021.3100486. <Go to ISI>://WOS:000683992800001.

Li, Xiaoyu, Jianhua Xu, Xuejing Ding, and Hongqiang Lyu. 2023. "State of Charge Estimation for Batteries Based on Common Feature Extraction and Transfer Learning." *Batteries-Basel* 9 (5). https://doi.org/10.3390/batteries9050266. <Go to ISI>://WOS:000995672100001.

Li, Yongkun, Chuang Wei, Yumao Sheng, Feipeng Jiao, and Kai Wu. 2020. "Swelling Force in Lithium-Ion Power Batteries." *Industrial & Engineering Chemistry Research* 59 (27): 12313–12318. https://doi.org/10.1021/acs.iecr.0c01035. <Go to ISI>://WOS:000550637000004.

Li, Yubai, Zhifu Zhou, and Wei-Tao Wu. 2020. "Three-Dimensional Thermal Modeling of Internal Shorting Process in a 20Ah Lithium-Ion Polymer Battery." *Energies* 13 (4). https://doi.org/10.3390/en13041013. <Go to ISI>://WOS:000522492700241.

Li, Yue, Ying-de Huang, Jing-yi Li, Chang-long Lei, Zhen-jiang He, Yi Cheng, Fei-xiang Wu, and Yun-jiao Li. 2023. "B-Doped Nickel-Rich Ternary Cathode Material for Lithium-Ion Batteries with Excellent Rate Performance." *Ionics*. https://doi.org/10.1007/s11581-023-05191-9. <Go to ISI>://WOS:001061902700005.

Lian, Gaoqi, Min Ye, Qiao Wang, Meng Wei, and Yuchuan Ma. 2023. "Noise-Immune State of Charge Estimation for Lithium-Ion Batteries Based on Optimized Dynamic Model and Improved Adaptive Unscented Kalman Filter Under Wide Temperature Range." *Journal of Energy Storage* 64. https://doi.org/10.1016/j.est.2023.107223. <Go to ISI>://WOS:000975594300001.

Liang, Shuang, Zhiyi Fang, Geng Sun, Chi Lin, Jiahui Li, Songyang Li, and Aimin Wang. 2021. "Charging UAV Deployment for Improving Charging Performance of Wireless Rechargeable Sensor Networks Via Joint Optimization Approach." *Computer Networks* 201. https://doi.org/10.1016/j.comnet.2021.108573. <Go to ISI>://WOS:000759699300017.

Lin, Mingqiang, Xianping Zeng, and Ji Wu. 2021. "State of Health Estimation of Lithium-Ion Battery Based on an Adaptive Tunable Hybrid Radial Basis Function Network." *Journal of Power Sources* 504. https://doi.org/10.1016/j.jpowsour.2021.230063. <Go to ISI>://WOS:000663408500001.

Lin, Qiongbin, Huasen Li, Qinqin Chai, Fenghuang Cai, and Yin Zhan. 2022. "Simultaneous and Rapid Estimation of State of Health and State of Charge for Lithium-Ion Battery Based on Response Characteristics of Load Surges." *Journal of Energy Storage* 55. https://doi.org/10.1016/j.est.2022.105495. <Go to ISI>://WOS:000862789700003.

Lin, Qiongbin, Zhifan Xu, and Chih-Min Lin. 2021. "State of Health Estimation and Remaining Useful Life Prediction for Lithium-Ion Batteries Using FBELNN and RCMNN." *Journal of Intelligent & Fuzzy Systems* 40 (6): 10919–10933. https://doi.org/10.3233/jifs-201952. <Go to ISI>://WOS:000667508800040.

Lin, Zhicheng, Houpeng Hu, Wei Liu, Zixia Zhang, Ya Zhang, Nankun Geng, and Qiangqiang Liao. 2023. "State of Health Estimation of Lithium-Ion Batteries Based on Remaining Area Capacity." *Journal of Energy Storage* 63. https://doi.org/10.1016/j.est.2023.107078. <Go to ISI>://WOS:000956577500001.

Ling, Liuyi, and Ying Wei. 2021. "State-of-Charge and State-of-Health Estimation for Lithium-Ion Batteries Based on Dual Fractional-Order Extended Kalman Filter and Online Parameter Identification." *IEEE Access* 9: 47588–47602. https://doi.org/10.1109/access.2021.3068813. <Go to ISI>://WOS:000637167900001.

Lipu, M. S. Hossain, M. A. Hannan, Aini Hussain, Afida Ayob, Mohamad H. M. Saad, and Kashem M. Muttaqi. 2020. "State of Charge Estimation in Lithium-Ion Batteries: A Neural Network Optimization Approach." *Electronics* 9 (9). https://doi.org/10.3390/electronics9091546. <Go to ISI>://WOS:000580850600001.

Liu, Boyang. 2020. *Electrochemical Model-based State Estimation and Parameter Identification for Lithium-ion Batteries*. Edited by Boyang Liu. Ann Arbor: ProQuest Dissertations Publishing.

Liu, Cuicui, Xiankui Wen, Jingliang Zhong, Wei Liu, Jianhong Chen, Jiawei Zhang, Zhiqin Wang, and Qiangqiang Liao. 2022. "Characterization of Aging Mechanisms and State of Health for Second-Life 21700 Ternary Lithium-Ion Battery." *Journal of Energy Storage* 55. https://doi.org/10.1016/j.est.2022.105511. <Go to ISI>://WOS:000874963400004.

Liu, Fang, Yan-peng Liu, Wei-xing Su, Chang-ping Jiao, and Yang Liu. 2021. "Online Estimation of Lithium-Ion Batteries State of Health During Discharge." *International Journal of Energy Research* 45 (7): 10112–10128. https://doi.org/10.1002/er.6502. <Go to ISI>://WOS:000616472800001.

Liu, Fang, Dan Yu, Weixing Su, and Fantao Bu. 2023. "Multi-State Joint Estimation of Series Battery Pack Based on Multi-Model Fusion." *Electrochimica Acta* 443. https://doi.org/10.1016/j.electacta.2023.141964. <Go to ISI>://WOS:000934314900001.

Liu, Hui, Fei Hu, Jinshuo Su, Xiaowei Wei, and Risheng Qin. 2020. "Comparisons on Kalman-Filter-Based Dynamic State Estimation Algorithms of Power Systems." *IEEE Access* 8: 51035–51043. https://doi.org/10.1109/access.2020.2979735. <Go to ISI>://WOS:000524899100005.

Liu, Jialong, Qiangling Duan, Kaixuan Qi, Yujun Liu, Jinhua Sun, Zhirong Wang, and Qingsong Wang. 2022. "Capacity Fading Mechanisms and State of Health Prediction of Commercial Lithium-Ion Battery in Total Lifespan."

*Journal of Energy Storage* 46. https://doi.org/10.1016/j.est.2021.103910. <Go to ISI>://WOS:000780239200002.

Liu, Kailong, Xiaopeng Tang, Remus Teodorescu, Furong Gao, and Jinhao Meng. 2022. "Future Ageing Trajectory Prediction for Lithium-Ion Battery Considering the Knee Point Effect." *IEEE Transactions on Energy Conversion* 37 (2): 1282–1291. https://doi.org/10.1109/tec.2021.3130600. <Go to ISI>://WOS:000800225300055.

Liu, Ke, Shunli Wang, Chunmei Yu, Chuangshi Qi, Xiao Yang, and Jialu Qiao. 2022. "A Novel Bias Compensated Recursive Least Squares and Multi Innovation Unscented Kalman Filtering Algorithm Method for Accurate State of Charge Estimation of Lithium-Ion Batteries." *International Journal of Electrochemical Science* 17 (9). https://doi.org/10.20964/2022.09.24. <Go to ISI>://WOS:000884756600018.

Liu, Lei, Chunjing Lin, Bin Fan, Fang Wang, Li Lao, and Peixia Yang. 2020. "A New Method to Determine the Heating Power of Ternary Cylindrical Lithium Ion Batteries With Highly Repeatable Thermal Runaway Test Characteristics." *Journal of Power Sources* 472. https://doi.org/10.1016/j.jpowsour.2020.228503. <Go to ISI>://WOS:000558890300026.

Liu, Lihua, Jianguo Zhu, and Linfeng Zheng. 2020. "An Effective Method for Estimating State of Charge of Lithium-Ion Batteries Based on an Electrochemical Model and Nernst Equation." *IEEE Access* 8: 211738–211749. https://doi.org/10.1109/access.2020.3039783. <Go to ISI>://WOS:000597217600001.

Liu, Shulin, Xia Dong, Xiaodong Yu, Xiaoqing Ren, Jinfeng Zhang, and Rui Zhu. 2022. "A Method for State of Charge and State of Health Estimation of Lithium-Ion Battery Based on Adaptive Unscented Kalman Filter." *Energy Reports* 8: 426–436. https://doi.org/10.1016/j.egyr.2022.09.093. <Go to ISI>://WOS:000897911400015.

Liu, Xiaoming, Hai Hu, Yaohong Suo, and Pengfei Yu. 2020. "Analysis of Diffusion Induced Deformation Considering Electro-Migration in Lithium-Ion Batteries." *International Journal of Electrochemical Science* 15 (7): 6012–6023. https://doi.org/10.20964/2020.07.31. <Go to ISI>://WOS:000551732800005.

Liu, Xingtao, Qiule Li, Li Wang, Mingqiang Lin, and Ji Wu. 2023. "Data-Driven State of Charge Estimation for Power Battery With Improved Extended Kalman Filter." *IEEE Transactions on Instrumentation and Measurement* 72. https://doi.org/10.1109/tim.2023.3239629. <Go to ISI>://WOS:000966981800001.

Liu, Xingtao, Chaoyi Zheng, Ji Wu, Jinhao Meng, Daniel-Ioan Stroe, and Jiajia Chen. 2020. "An Improved State of Charge and State of Power Estimation Method Based on Genetic Particle Filter for Lithium-ion Batteries." *Energies* 13 (2). https://doi.org/10.3390/en13020478. <Go to ISI>://WOS:000520432300192.

Liu, Xintian, Xuhui Deng, Yao He, Xinxin Zheng, and Guojian Zeng. 2020. "A Dynamic State-of-Charge Estimation Method for Electric Vehicle Lithium-Ion Batteries." *Energies* 13 (1). https://doi.org/10.3390/en13010121. <Go to ISI>://WOS:000520425800121.

Liu, Yan, Weiguang Lv, Xiaohong Zheng, Dingshan Ruan, Yongxia Yang, Hongbin Cao, and Zhi Sun. 2021. "Near-to-Stoichiometric Acidic Recovery of Spent Lithium-Ion Batteries through Induced Crystallization." *ACS Sustainable Chemistry & Engineering* 9 (8): 3183–3194. https://doi.org/10.1021/acssuschemeng.0c08163. <Go to ISI>://WOS:000625460400017.

Liu, Yiqun. 2022. *Pouch-Type $LiNiMnCoO_2$ Battery Experiments and Simulation for Electrified Light-Duty Vehicles.* Edited by Yiqun Liu. Ann Arbor: ProQuest Dissertations Publishing.

Liu, Yuefeng, Yingjie He, Haodong Bian, Wei Guo, and Xiaoyan Zhang. 2022. "A Review of Lithium-Ion Battery State of Charge Estimation Based on Deep Learning: Directions for Improvement and Future Trends." *Journal of Energy Storage* 52. https://doi.org/10.1016/j.est.2022.104664. <Go to ISI>://WOS:000804763300005.

Liu, Yuefeng, Jiaqi Li, Gong Zhang, Bin Hua, and Neal Xiong. 2021. "State of Charge Estimation of Lithium-Ion Batteries Based on Temporal Convolutional Network and Transfer Learning." *IEEE Access* 9: 34177–34187. https://doi.org/10.1109/access.2021.3057371. <Go to ISI>://WOS:000626309500001.

Liu, Yuyang, Shunli Wang, Yanxin Xie, Carlos Fernandez, Jingsong Qiu, and Yixing Zhang. 2022. "A Novel Adaptive H-Infinity Filtering Method for the Accurate SOC Estimation of Lithium-Ion Batteries Based on Optimal Forgetting Factor Selection." *International Journal of Circuit Theory and Applications* 50 (10): 3372–3386. https://doi.org/10.1002/cta.3339. <Go to ISI>://WOS:000805660200001.

Liu, Zhiqiang, Weidong Wang, Junyi He, Jianjun Zhang, Jing Wang, Shasha Li, Yining Sun, and Xianyang Ren. 2023. "A New Hybrid Algorithm for Vehicle Routing Optimization." *Sustainability* 15 (14). https://doi.org/10.3390/su151410982. <Go to ISI>://WOS:001071411600001.

Liu, Zongwei, Xinglong Liu, Han Hao, Fuquan Zhao, Amer Ahmad Amer, and Hassan Babiker. 2020. "Research on the Critical Issues for Power Battery Reusing of New Energy Vehicles in China." *Energies* 13 (8). https://doi.org/10.3390/en13081932. <Go to ISI>://WOS:000538041800075.

Lou, Yangbing. 2021. *Hybrid Approaches of Battery Performance Modeling and Prognosis.* Edited by Yangbing Lou. Ann Arbor: ProQuest Dissertations Publishing.

Lu, Mengyue, Weiwei Yang, Yiming Deng, and Qian Xu. 2020. "An Optimal Electrolyte Addition Strategy for Improving Performance of a Vanadium Redox Flow Battery." *International Journal of Energy Research* 44 (4): 2604–2616. https://doi.org/10.1002/er.4988. <Go to ISI>://WOS:000507160000001.

Luciani, Sara, Stefano Feraco, Angelo Bonfitto, and Andrea Tonoli. 2021. "Hardware-in-the-Loop Assessment of a Data-Driven State of Charge Estimation Method for Lithium-Ion Batteries in Hybrid Vehicles." *Electronics* 10 (22). https://doi.org/10.3390/electronics10222828. <Go to ISI>://WOS:000723911300001.

Luo, Wei, Adnan U. Syed, John R. Nicholls, and Simon Gray. 2023. "An SVM-Based Health Classifier for Offline Li-Ion Batteries by Using EIS Technology."

*Journal of the Electrochemical Society* 170 (3). https://doi.org/10.1149/1945-7111/acc09f. <Go to ISI>://WOS:000955738100001.

Lv, Haichao, Xiankun Huang, and Yongzhong Liu. 2020. "Analysis on Pulse Charging-Discharging Strategies for Improving Capacity Retention Rates of Lithium-Ion Batteries." *Ionics* 26 (4): 1749–1770. https://doi.org/10.1007/s11581-019-03404-8. <Go to ISI>://WOS:000507692000001.

L'Vov, P. E., M. Yu Tikhonchev, and R. T. Sibatov. 2022. "Phase-Field Model of Ion Transport and Intercalation in Lithium-Ion Battery." *Journal of Energy Storage* 50. https://doi.org/10.1016/j.est.2022.104319. <Go to ISI>://WOS:000780272400006.

Ma, Yan, Meihao Yao, Hongcheng Liu, and Zhiguo Tang. 2022. "State of Health Estimation and Remaining Useful Life Prediction for Lithium-Ion Batteries by Improved Particle Swarm Optimization-Back Propagation Neural Network." *Journal of Energy Storage* 52. https://doi.org/10.1016/j.est.2022.104750. <Go to ISI>://WOS:000798298000006.

Makeen, Peter, Hani A. Ghali, and Saim Memon. 2020. "Experimental and Theoretical Analysis of the Fast Charging Polymer Lithium-Ion Battery Based on Cuckoo Optimization Algorithm (COA)." *IEEE Access* 8: 140486–140496. https://doi.org/10.1109/access.2020.3012913. <Go to ISI>://WOS:000560580200001.

Manukumar, K. N., Brij Kishore, R. Viswanatha, and G. Nagaraju. 2020. "$Ta_2O_5$ Nanoparticles as an Anode Material for Lithium Ion Battery." *Journal of Solid State Electrochemistry* 24 (4): 1067–1074. https://doi.org/10.1007/s10008-020-04593-3. <Go to ISI>://WOS:000526269000004.

Mao, Jia, Dou Hong, Runwang Ren, and Xiangyu Li. 2020. "The Effect of Marine Power Generation Technology on the Evolution of Energy Demand for New Energy Vehicles." *Journal of Coastal Research*: 1006–1009. https://doi.org/10.2112/si103-209.1. <Go to ISI>://WOS:000543720600209.

Mao, Ling, Qin Hu, Jinbin Zhao, and Xiaofang Yu. 2023. "State-of-Charge of Lithium-Ion Battery Based on Equivalent Circuit Model – Relevance Vector Machine Fusion Model Considering Varying Ambient Temperatures." *Measurement* 221. https://doi.org/10.1016/j.measurement.2023.113487. <Go to ISI>://WOS:001073934400001.

Mao, Ling, Huizhong Hu, Jiajun Chen, Jinbin Zhao, Keqing Qu, and Lei Jiang. 2023. "Online State-of-Health Estimation Method for Lithium-Ion Battery Based on CEEMDAN for Feature Analysis and RBF Neural Network." *IEEE Journal of Emerging and Selected Topics in Power Electronics* 11 (1): 187–200. https://doi.org/10.1109/jestpe.2021.3106708. <Go to ISI>://WOS:000966602700001.

Mao, Ling, Chuan Yang, Jinbin Zhao, Keqing Qu, and Xiaofang Yu. 2023. "Joint State-of-Charge and State-of-Health Estimation for Lithium-Ion Batteries Based on Improved Lebesgue Sampling and Division of Aging Stage." *Energy Technology*. https://doi.org/10.1002/ente.202300567. <Go to ISI>://WOS:001048656700001.

Mashhour, Emad Mohamed, Enas M. F. El Houby, Khaled Tawfik Wassif, and Akram Ibrahim Salah. 2020. "A Novel Classifier Based on Firefly

Algorithm." *Journal of King Saud University-Computer and Information Sciences* 32 (10): 1173–1181. https://doi.org/10.1016/j.jksuci.2018.11.009. <Go to ISI>://WOS:000603358200008.

Meng, Huixing, Mengyao Geng, and Te Han. 2023. "Long Short-Term Memory Network with Bayesian Optimization for Health Prognostics of Lithium-Ion Batteries Based on Partial Incremental Capacity Analysis." *Reliability Engineering & System Safety* 236. https://doi.org/10.1016/j.ress.2023.109288. <Go to ISI>://WOS:000983499500001.

Menz, Fabian, Marius Bauer, Olaf Boese, Moritz Pausch, and Michael A. Danzer. 2023. "Investigating the Thermal Runaway Behaviour of Fresh and Aged Large Prismatic Lithium-Ion Cells in Overtemperature Experiments." *Batteries-Basel* 9 (3). https://doi.org/10.3390/batteries9030159. <Go to ISI>://WOS:000954116300001.

Miao, Youping, Lili Liu, Kaihua Xu, and Jinhui Li. 2023. "High Concentration from Resources to Market Heightens Risk for Power Lithium-Ion Battery Supply Chains Globally." *Environmental Science and Pollution Research* 30 (24): 65558–65571. https://doi.org/10.1007/s11356-023-27035-9. <Go to ISI>://WOS:000975912800003.

Michalski, Jacek, Piotr Kozierski, Wojciech Giernacki, Joanna Zietkiewicz, and Marek Retinger. 2021. "MultiPDF Particle Filtering in State Estimation of Nonlinear Objects." *Nonlinear Dynamics* 106 (3): 2165–2182. https://doi.org/10.1007/s11071-021-06913-2. <Go to ISI>://WOS:000710875400003.

Migallon, Hector, Akram Belazi, Jose-Luis Sanchez-Romero, Hector Rico, and Antonio Jimeno-Morenilla. 2020. "Settings-Free Hybrid Metaheuristic General Optimization Methods." *Mathematics* 8 (7). https://doi.org/10.3390/math8071092. <Go to ISI>://WOS:000556744700001.

Migallon, Hector, A. Jimeno-Morenilla, J. L. Sanchez-Romero, and A. Belazi. 2020. "Efficient Parallel and Fast Convergence Chaotic Jaya Algorithms." *Swarm and Evolutionary Computation* 56. https://doi.org/10.1016/j.swevo.2020.100698. <Go to ISI>://WOS:000540614100003.

Mohammed, Nabil, and Ahmed Majed Saif. 2021. "Programmable Logic Controller Based Lithium-Ion Battery Management System for Accurate State of Charge Estimation." *Computers & Electrical Engineering* 93. https://doi.org/10.1016/j.compeleceng.2021.107306. <Go to ISI>://WOS:000687736100071.

Mohan, Indra, Anshu Raj, Kumar Shubham, D. B. Lata, Sandip Mandal, and Sachin Kumar. 2022. "Potential of Potassium and Sodium-Ion Batteries as the Future of Energy Storage: Recent Progress in Anodic Materials." *Journal of Energy Storage* 55. https://doi.org/10.1016/j.est.2022.105625. <Go to ISI>://WOS:000911762700004.

Mohanty, Smita, and Rajashree Dash. 2023. "RVFLN-CDFPA: A Random Vector Functional Link Neural Network Optimized Using a Chaotic Differential Flower Pollination Algorithm for Day Ahead Net Asset Value Prediction." *Evolving Systems*. https://doi.org/10.1007/s12530-023-09501-4. <Go to ISI>://WOS:001029553700001.

Na, Ying, Xiaohong Sun, Anran Fan, Shu Cai, and Chunming Zheng. 2021. "Methods for Enhancing the Capacity of Electrode Materials in

Low-Temperature Lithium-Ion Batteries." *Chinese Chemical Letters* 32 (3): 973–982. https://doi.org/10.1016/j.cclet.2020.09.007. <Go to ISI>://WOS:000642405200003.

Natella, Domenico, Simona Onori, and Francesco Vasca. 2023. "A Co-Estimation Framework for State of Charge and Parameters of Lithium-Ion Battery With Robustness to Aging and Usage Conditions." *IEEE Transactions on Industrial Electronics* 70 (6): 5760–5770. https://doi.org/10.1109/tie.2022.3194576. <Go to ISI>://WOS:000965350800001.

Navega Vieira, Romulo, Juan Moises Mauricio Villanueva, Thommas Kevin Sales Flores, and Euler Cassio Tavares de Macedo. 2022. "State of Charge Estimation of Battery Based on Neural Networks and Adaptive Strategies with Correntropy." *Sensors* 22 (3). https://doi.org/10.3390/s22031179. <Go to ISI>://WOS:000755998600001.

Obeid, Hussein, Raffaele Petrone, Hicham Chaoui, and Hamid Gualous. 2023. "Higher Order Sliding-Mode Observers for State-of-Charge and State-of-Health Estimation of Lithium-Ion Batteries." *IEEE Transactions on Vehicular Technology* 72 (4): 4482–4492. https://doi.org/10.1109/tvt.2022.3226686. <Go to ISI>://WOS:000975101300027.

Osara, Jude A., Ofodike A. Ezekoye, Kevin C. Marr, and Michael D. Bryant. 2021. "A Methodology for Analyzing aging and Performance of Lithium-Ion Batteries: Consistent Cycling Application." *Journal of Energy Storage* 42. https://doi.org/10.1016/j.est.2021.103119. <Go to ISI>://WOS:000701736200001.

Ould Ely, Teyeb, Dana Kamzabek, and Dhritiman Chakraborty. 2019. "Batteries Safety: Recent Progress and Current Challenges." *Frontiers in Energy Research* 7. https://doi.org/10.3389/fenrg.2019.00071. <Go to ISI>://WOS:000496930600001.

Ouyang, Dongxu, Jingwen Weng, Mingyi Chen, Jian Wang, and Zhirong Wang. 2022. "Electrochemical and Thermal Characteristics of Aging Lithium-Ion Cells After Long-Term Cycling at Abusive-Temperature Environments." *Process Safety and Environmental Protection* 159: 1215–1223. https://doi.org/10.1016/j.psep.2022.01.055. <Go to ISI>://WOS:000781725000003.

Ouyang, Dongxu, Jingwen Weng, Mingyi Chen, Jian Wang, and Zhirong Wang. 2023. "Investigation on Topographic, Electrochemical and Thermal Features of Aging Lithium-Ion Cells Induced by Overcharge/Over-Discharge Cycling." *Journal of Energy Storage* 68. https://doi.org/10.1016/j.est.2023.107799. <Go to ISI>://WOS:001013491500001.

Ouyang, Nan, Wencan Zhang, Xiuxing Yin, Xingyao Li, Yi Xie, Hancheng He, and Zhuoru Long. 2023. "A Data-Driven Method for Predicting Thermal Runaway Propagation of Battery Modules Considering Uncertain Conditions." *Energy* 273. https://doi.org/10.1016/j.energy.2023.127168. <Go to ISI>://WOS:000959328600001.

Ouyang, Quan, Jian Chen, and Jian Zheng. 2020. "State-of-Charge Observer Design for Batteries With Online Model Parameter Identification: A Robust Approach." *IEEE Transactions on Power Electronics* 35 (6): 5820–5831. https://doi.org/10.1109/tpel.2019.2948253. <Go to ISI>://WOS:000554997600027.

Ouyang, Tiancheng, Peihang Xu, Jie Lu, Xiaoyi Hu, Benlong Liu, and Nan Chen. 2022. "Coestimation of State-of-Charge and State-of-Health for Power Batteries Based on Multithread Dynamic Optimization Method." *IEEE Transactions on Industrial Electronics* 69 (2): 1157–1166. https://doi.org/10.1109/tie.2021.3062266. <Go to ISI>://WOS:000712582800012.

Pai, Hung Yu, Yi Hua Liu, and Song Pei Ye. 2023. "Online Estimation of Lithium-Ion Battery Equivalent Circuit Model Parameters and State of Charge Using Time-Domain Assisted Decoupled Recursive Least Squares Technique." *Journal of Energy Storage* 62. https://doi.org/10.1016/j.est.2023.106901. <Go to ISI>://WOS:000946263100001.

Pan, Bin, Dong Dong, Jionggeng Wang, Jianbo Nie, Shuangyu Liu, Yaohe Cao, and Yinzhu Jiang. 2020. "Aging Mechanism Diagnosis of Lithium Ion Battery by Open Circuit Voltage Analysis." *Electrochimica Acta* 362. https://doi.org/10.1016/j.electacta.2020.137101. <Go to ISI>://WOS:000582869700016.

Pan, Ting-Chen, En-Jui Liu, Hung-Chih Ku, and Che-Wun Hong. 2022. "Parameter Identification and Sensitivity Analysis of Lithium-Ion Battery Via Whale Optimization Algorithm." *Electrochimica Acta* 404. https://doi.org/10.1016/j.electacta.2021.139574. <Go to ISI>://WOS:000779416300012.

Pan, Yen-ting, and Yonhua Tzeng. 2019. "Silicon Nanoparticles in Graphene Sponge for Long-Cycling-Life and High-Capacity Anode of Lithium Ion Battery." *IEEE Transactions on Nanotechnology* 18: 1097–1102. https://doi.org/10.1109/tnano.2019.2946459. <Go to ISI>://WOS:000494426000002.

Pang, Hui, Yuanfei Geng, Xiaofei Liu, and Longxing Wu. 2022. "A Composite State of Charge Estimation for Electric Vehicle Lithium-Ion Batteries Using Back-Propagation Neural Network and Extended Kalman Particle Filter." *Journal of the Electrochemical Society* 169 (11). https://doi.org/10.1149/1945-7111/ac9f79. <Go to ISI>://WOS:000883764600001.

Pang, Hui, Jiamin Jin, Longxing Wu, Fengqi Zhang, and Kai Liu. 2021. "A Comprehensive Physics-Based Equivalent-Circuit Model and State of Charge Estimation for Lithium-Ion Batteries." *Journal of the Electrochemical Society* 168 (9). https://doi.org/10.1149/1945-7111/ac2701. <Go to ISI>://WOS:000698815300001.

Park, Sang-Jun, Young-Woong Song, Byeong-Su Kang, Woo-Joong Kim, Yeong-Jun Choi, Chanhoon Kim, and Young-Sun Hong. 2023. "Depth of Discharge Characteristics and Control Strategy to Optimize Electric Vehicle Battery Life." *Journal of Energy Storage* 59. https://doi.org/10.1016/j.est.2022.106477. <Go to ISI>://WOS:000918776300001.

Park, Seongyun, Jonghoon Kim, and Inho Cho. 2023. "Hybrid SOC and SOH Estimation Method With Improved Noise Immunity and Computational Efficiency in Hybrid Railroad Propulsion System." *Journal of Energy Storage* 72. https://doi.org/10.1016/j.est.2023.108385. <Go to ISI>://WOS:001050120900001.

Park, Seong Yun, Pyeong Yeon Lee, Yoo, Kisoo, and Jong Hoon Kim. 2021. "A SOH Estimation Study on Lithium-Ion Battery based on Incremental Capacity and Differential Voltage Analysis." *Transactions of the KSME, A* 45 (3): 259–266. <Go to ISI>://KJD:ART002686722.

Park, Sun Woo, Hyunju Lee, and Yong Sul Won. 2022. "A Novel Aging Parameter Method for Online Estimation of Lithium-Ion Battery States of Charge and Health." *Journal of Energy Storage* 48. https://doi.org/10.1016/j.est.2022.103987. <Go to ISI>://WOS:000780269200002.

Paul, Theophile, Tedjani Mesbahi, Sylvain Durand, Damien Flieller, and Wilfried Uhring. 2020. "Sizing of Lithium-Ion Battery/Supercapacitor Hybrid Energy Storage System for Forklift Vehicle." *Energies* 13 (17). https://doi.org/10.3390/en13174518. <Go to ISI>://WOS:000569975000001.

Peng, Jichang, Jinhao Meng, Dan Chen, Haitao Liu, Sipeng Hao, Xin Sui, and Xinghao Du. 2022. "A Review of Lithium-Ion Battery Capacity Estimation Methods for Onboard Battery Management Systems: Recent Progress and Perspectives." *Batteries-Basel* 8 (11). https://doi.org/10.3390/batteries8110229. <Go to ISI>://WOS:000894311900001.

Pierezan, Juliano, Leandro dos Santos Coelho, Viviana Cocco Mariani, Emerson Hochsteiner de Vasconcelos Segundo, and Doddy Prayogo. 2021. "Chaotic Coyote Algorithm Applied to Truss Optimization Problems." *Computers & Structures* 242. https://doi.org/10.1016/j.compstruc.2020.106353. <Go to ISI>://WOS:000580652800001.

Porter, Stephen R. 2021. "Point Set Registration Via Stochastic Particle Flow Filter." *Journal of Electronic Imaging* 30 (6). https://doi.org/10.1117/1.Jei.30.6.063007. <Go to ISI>://WOS:000737383500007.

Potrykus, Szymon, Filip Kutt, Janusz Nieznanski, and Francisco Jesus Fernandez Morales. 2020. "Advanced Lithium-Ion Battery Model for Power System Performance Analysis." *Energies* 13 (10). https://doi.org/10.3390/en13102411. <Go to ISI>://WOS:000539257300004.

Premkumar, M., R. Sowmya, S. Sridhar, C. Kumar, Mohamed Abbas, Malak S. Alqahtani, and Kottakkaran Sooppy Nisar. 2022. "State-of-Charge Estimation of Lithium-Ion Battery for Electric Vehicles Using Deep Neural Network." *Cmc-Computers Materials & Continua* 73 (3): 6289–6306. https://doi.org/10.32604/cmc.2022.030490. <Go to ISI>://WOS:000864725100033.

Qi, Chuangshi, Shunli Wang, Wen Cao, Haotian Shi, and Yanxin Xie. 2022. "A Novel Multi-Constraint Peak Power Prediction Method Combined With Online Model Parameter Identification and State-of-Charge Co-Estimation of Lithium-Ion Batteries." *Journal of the Electrochemical Society* 169 (12). https://doi.org/10.1149/1945-7111/aca721. <Go to ISI>://WOS:000894212000001.

Qian, Kangfeng, Xintian Liu, Yiquan Wang, Xueguang Yu, and Bixiong Huang. 2022. "Modified Dual Extended Kalman Filters for SOC Estimation and Online Parameter Identification of Lithium-Ion Battery Via Modified Gray Wolf Optimizer." *Proceedings of the Institution of Mechanical Engineers Part D-Journal of Automobile Engineering* 236 (8): 1761–1774. https://doi.org/10.1177/09544070211046693. <Go to ISI>://WOS:000695300300001.

Qiao, Jialu, Shunli Wang, Chunmei Yu, Weihao Shi, and Carlos Fernandez. 2021. "A Novel Bias Compensation Recursive Least Square-Multiple Weighted Dual Extended Kalman Filtering Method for Accurate State-of-Charge and State-of-Health Co-Estimation of Lithium-Ion Batteries." *International*

*Journal of Circuit Theory and Applications* 49 (11): 3879–3893. https://doi.org/10.1002/cta.3115. <Go to ISI>://WOS:000686387500001.

Qiao, Junhao, Jingping Liu, Jichao Liang, Dongdong Jia, Rumin Wang, Dazi Shen, and Xiongbo Duan. 2023. "Experimental Investigation the Effects of Miller Cycle Coupled With Asynchronous Intake Valves on Cycle-to-Cycle Variations and Performance of the SI Engine." *Energy* 263. https://doi.org/10.1016/j.energy.2022.125868. <Go to ISI>://WOS:000882431400001.

Qiu, Xianghui, Weixiong Wu, and Shuangfeng Wang. 2020. "Remaining Useful Life Prediction of Lithium-Ion Battery Based on Improved Cuckoo Search Particle Filter and a Novel State of Charge Estimation Method." *Journal of Power Sources* 450. https://doi.org/10.1016/j.jpowsour.2020.227700. <Go to ISI>://WOS:000517663800037.

Qu, Jiantao, Feng Liu, Yuxiang Ma, and Jiaming Fan. 2019. "A Neural-Network-Based Method for RUL Prediction and SOH Monitoring of Lithium-Ion Battery." *IEEE Access* 7: 87178–87191. https://doi.org/10.1109/access.2019.2925468. <Go to ISI>://WOS:000476874200001.

Ragone, Marco, Vitaliy Yurkiv, Ajaykrishna Ramasubramanian, Babak Kashir, and Farzad Mashayek. 2021. "Data Driven Estimation of Electric Vehicle Battery State-of-Charge Informed by Automotive Simulations and Multi-Physics Modeling." *Journal of Power Sources* 483. https://doi.org/10.1016/j.jpowsour.2020.229108. <Go to ISI>://WOS:000621297200001.

Rahimifard, Sara, Saeid Habibi, Gillian Goward, and Jimi Tjong. 2021. "Adaptive Smooth Variable Structure Filter Strategy for State Estimation of Electric Vehicle Batteries." *Energies* 14 (24). https://doi.org/10.3390/en14248560. <Go to ISI>://WOS:000739156500001.

Raijmakers, L. H. J., D. L. Danilov, R. A. Eichel, and P. H. L. Notten. 2020. "An Advanced All-Solid-State Li-Ion Battery Model." *Electrochimica Acta* 330. https://doi.org/10.1016/j.electacta.2019.135147. <Go to ISI>://WOS:000501468400002.

Redondo-Iglesias, Eduardo, Pascal Venet, and Serge Pelissier. 2020. "Modelling Lithium-Ion Battery Ageing in Electric Vehicle Applications-Calendar and Cycling Ageing Combination Effects." *Batteries-Basel* 6 (1). https://doi.org/10.3390/batteries6010014. <Go to ISI>://WOS:000523703600014.

Ren, Dongsheng, Xuning Feng, Lishuo Liu, Hungjen Hsu, Languang Lu, Li Wang, Xiangming He, and Minggao Ouyang. 2021. "Investigating the Relationship Between Internal Short Circuit and Thermal Runaway of Lithium-Ion Batteries Under Thermal Abuse Condition." *Energy Storage Materials* 34: 563–573. https://doi.org/10.1016/j.ensm.2020.10.020. <Go to ISI>://WOS:000598776100002.

Ren, Pu, Shunli Wang, Junhan Huang, Xianpei Chen, Mingfang He, and Wen Cao. 2022. "Novel Co-Estimation Strategy Based on Forgetting Factor Dual Particle Filter Algorithm for the State of Charge and State of Health of the Lithium-Ion Battery." *International Journal of Energy Research* 46 (2): 1094–1107. https://doi.org/10.1002/er.7230. <Go to ISI>://WOS:000694935700001.

Ren, Wenju, Taixiong Zheng, Changhao Piao, Daryn Eugene Benson, Xin Wang, Haiqing Li, and Shen Lu. 2022. "Characterization of Commercial 18,650

Li-Ion Batteries Using Strain Gauges." *Journal of Materials Science* 57 (28): 13560–13569. https://doi.org/10.1007/s10853-022-07490-4. <Go to ISI>://WOS:000826851600003.

Ren, Xiangyang, Shuai Chen, Kunyuan Wang, and Juan Tan. 2022. "Design and Application of Improved Sparrow Search Algorithm Based on Sine Cosine and Firefly Perturbation." *Mathematical Biosciences and Engineering* 19 (11): 11422–11452. https://doi.org/10.3934/mbe.2022533. <Go to ISI>://WOS:000841170200002.

Rezaei, Kamran, and Hassan Rezaei. 2022. "An Improved Firefly Algorithm for Numerical Optimization Problems and It's Application in Constrained Optimization." *Engineering with Computers* 38 (4): 3793–3813. https://doi.org/10.1007/s00366-021-01412-9. <Go to ISI>://WOS:000650315400002.

Rezaei, Omid, Reza Habibifar, and Zhanle Wang. 2022. "A Robust Kalman Filter-Based Approach for *SoC* Estimation of Lithium-Ion Batteries in Smart Homes." *Energies* 15 (10). https://doi.org/10.3390/en15103768. <Go to ISI>://WOS:000802648800001.

Rimsha, Sadia, Sadia Murawwat, Muhammad Majid Gulzar, Ahmad Alzahrani, Ghulam Hafeez, Farrukh Aslam Khan, and Azher M. Abed. 2023. "State of Charge Estimation and Error Analysis of Lithium-Ion Batteries for Electric Vehicles Using Kalman Filter and Deep Neural Network." *Journal of Energy Storage* 72. https://doi.org/10.1016/j.est.2023.108039. <Go to ISI>://WOS:001052275300001.

Rudyi, A. S., A. A. Mironenko, V. V. Naumov, A. M. Skundin, T. L. Kulova, I. S. Fedorov, and S. V. Vasil'ev. 2020. "A Solid-State Lithium-Ion Battery: Structure, Technology, and Characteristics." *Technical Physics Letters* 46 (3): 215–219. https://doi.org/10.1134/s1063785020030141. <Go to ISI>://WOS:000529352900004.

Rumberg, Bjoern, Kai Schwarzkopf, Bernd Epding, Ina Stradtmann, and Arno Kwade. 2019. "Understanding the Different Aging Trends of Usable Capacity and Mobile Li Capacity in Li-Ion Cells." *Journal of Energy Storage* 22: 336–344. https://doi.org/10.1016/j.est.2019.02.029. <Go to ISI>://WOS:000462506300036.

Saeed Alzaeemi, Shehab Abdulhabib, and Saratha Sathasivam. 2021. "Examining the Forecasting Movement of Palm Oil Price Using RBFNN-2SATRA Metaheuristic Algorithms for Logic Mining." *IEEE Access* 9: 22542–22557. https://doi.org/10.1109/access.2021.3054816. <Go to ISI>://WOS:000617372600001.

Sakile, Rajakumar, and Umesh Kumar Sinha. 2021. "Estimation of State of Charge and State of Health of Lithium-Ion Batteries Based on a New Adaptive Nonlinear Observer." *Advanced Theory and Simulations* 4 (11). https://doi.org/10.1002/adts.202100258. <Go to ISI>://WOS:000703320200001.

Sato, Kosuke, Akihiko Kono, Hiroaki Urushibata, Yoji Fujita, and Masato Koyama. 2019. "Physics-Based Model of Lithium-Ion Batteries Running on a Circuit Simulator." *Electrical Engineering in Japan* 208 (3–4): 48–63. https://doi.org/10.1002/eej.23237. <Go to ISI>://WOS:000480063800001.

Schlesinger, William H., Emily M. Klein, Zhen Wang, and Avner Vengosh. 2021. "Global Biogeochemical Cycle of Lithium." *Global Biogeochemical Cycles* 35 (8). https://doi.org/10.1029/2021gb006999. <Go to ISI>://WOS:000690773700003.

Shaban, Wafaa Mohamed, Khalid Elbaz, Mohamed Amin, and Ayat Gamal Ashour. 2022. "A New Systematic Firefly Algorithm for Forecasting the Durability of Reinforced Recycled Aggregate Concrete." *Frontiers of Structural and Civil Engineering* 16 (3): 329–346. https://doi.org/10.1007/s11709-022-0801-9. <Go to ISI>://WOS:000787416000003.

Shan, Haipeng, Hui Cao, Xing Xu, Teng Xiao, Guangya Hou, Huazhen Cao, Yiping Tang, and Guoqu Zheng. 2022. "Investigation of Self-Discharge Properties and a New Concept of Open-Circuit Voltage Drop Rate in Lithium-Ion Batteries." *Journal of Solid State Electrochemistry* 26 (1): 163–170. https://doi.org/10.1007/s10008-021-05049-y. <Go to ISI>://WOS:000702194700001.

Shang, Yunlong, Kailong Liu, Naxin Cui, Qi Zhang, and Chenghui Zhang. 2020. "A Sine-Wave Heating Circuit for Automotive Battery Self-Heating at Subzero Temperatures." *IEEE Transactions on Industrial Informatics* 16 (5): 3355–3365. https://doi.org/10.1109/tii.2019.2923446. <Go to ISI>://WOS:000519588700045.

Shen, Shuaiqi, Baochang Liu, Kuan Zhang, and Song Ci. 2021. "Toward Fast and Accurate SOH Prediction for Lithium-Ion Batteries." *IEEE Transactions on Energy Conversion* 36 (3): 2036–2046. https://doi.org/10.1109/tec.2021.3052504. <Go to ISI>://WOS:000681269600046.

Shen, Xianfeng, Shunli Wang, Chunmei Yu, Chuangshi Qi, Zehao Li, and Carlos Fernandez. 2023. "A Hybrid Algorithm Based on Beluga Whale Optimization-Forgetting Factor Recursive Least Square and Improved Particle Filter for the State of Charge Estimation of Lithium-Ion Batteries." *Ionics*. https://doi.org/10.1007/s11581-023-05147-z. <Go to ISI>://WOS:001048074800003.

Sheng, Hanmin, Yuan Zhou, Libing Bai, and Lei Shi. 2022. "Transfer State of Health Estimation Based on Cross-Manifold Embedding." *Journal of Energy Storage* 47. https://doi.org/10.1016/j.est.2021.103555. <Go to ISI>://WOS:000780266400004.

Shi, Junchuan, Alexis Rivera, and Dazhong Wu. 2022. "Battery Health Management Using Physics-Informed Machine Learning: Online Degradation Modeling and Remaining Useful Life Prediction." *Mechanical Systems and Signal Processing* 179. https://doi.org/10.1016/j.ymssp.2022.109347. <Go to ISI>://WOS:000833406400002.

Shi, Na, Zewang Chen, Mu Niu, Zhijia He, Youren Wang, and Jiang Cui. 2022. "State-of-Charge Estimation for the Lithium-Ion Battery Based on Adaptive Extended Kalman Filter Using Improved Parameter Identification." *Journal of Energy Storage* 45. https://doi.org/10.1016/j.est.2021.103518. <Go to ISI>://WOS:000779411200001.

Shin, Mi-Ra, Seon-Jin Lee, Seong-Jae Kim, and Tae-Whan Hong. 2020. "Preparation and Effect of (3-Aminopropyl)Triethoxysilane-Coated $LiNi_{0.5}Co_{0.2}Mn_{0.3}O_2$ Cathode Material for Lithium Ion Batteries." *Journal of*

*Nanoscience and Nanotechnology* 20 (6): 3460–3465. https://doi.org/10.1166/jnn.2020.17406. <Go to ISI>://WOS:000508548200016.

Shrivastava, Prashant, Tey Kok Soon, Mohd Yamani Idna Bin Idris, Saad Mekhilef, and Syed Bahari Ramadzan Syed Adnan. 2021. "Combined State of Charge and State of Energy Estimation of Lithium-Ion Battery Using Dual Forgetting Factor-Based Adaptive Extended Kalman Filter for Electric Vehicle Applications." *IEEE Transactions on Vehicular Technology* 70 (2): 1200–1215. https://doi.org/10.1109/tvt.2021.3051655. <Go to ISI>://WOS:000628913700011.

Smith, Jeffrey G., and Donald J. Siegel. 2020. "Low-Temperature Paddlewheel Effect in Glassy Solid Electrolytes." *Nature Communications* 11 (1). https://doi.org/10.1038/s41467-020-15245-5. <Go to ISI>://WOS:000522096100003.

Song, Yuchen, Yu Peng, and Datong Liu. 2021. "Model-Based Health Diagnosis for Lithium-Ion Battery Pack in Space Applications." *IEEE Transactions on Industrial Electronics* 68 (12): 12375–12384. https://doi.org/10.1109/tie.2020.3045745. <Go to ISI>://WOS:000692884200071.

Sonwane, AdityaNitin. 2020. *Mechanical Integrity of Cylindrical 21700 Lithium-Ion Batteries Based on Electrochemical Status*. Edited by AdityaNitin Sonwane. Ann Arbor: ProQuest Dissertations Publishing.

Stinson, Kerrek. 2023. "Existence for a Cahn-Hilliard Model for Lithium-Ion Batteries with Exponential Growth Boundary Conditions." *Journal of Nonlinear Science* 33 (5). https://doi.org/10.1007/s00332-023-09927-9. <Go to ISI>://WOS:001002747700001.

Su, Jie, Maosong Lin, Shunli Wang, Jin Li, James Coffie-Ken, and Fei Xie. 2019. "An Equivalent Circuit Model Analysis for the Lithium-Ion Battery Pack in Pure Electric Vehicles." *Measurement & Control* 52 (3–4): 193–201. https://doi.org/10.1177/0020294019827338. <Go to ISI>://WOS:000482917200005.

Sun, Bing, and Zhuofang Ju. 2023. "Research on the Promotion of New Energy Vehicles Based on Multi-Source Heterogeneous Data: Consumer and Manufacturer Perspectives." *Environmental Science and Pollution Research* 30 (11): 28863–28873. https://doi.org/10.1007/s11356-022-24304-x. <Go to ISI>://WOS:000885414700015.

Sun, Congkai, Xiong Zhang, Chen Li, Kai Wang, Xianzhong Sun, and Yanwei Ma. 2020. "High-Efficiency Sacrificial Prelithiation of Lithium-Ion Capacitors with Superior Energy-Storage Performance." *Energy Storage Materials* 24: 160–166. https://doi.org/10.1016/j.ensm.2019.08.023. <Go to ISI>://WOS:000500484000013.

Sun, Hanlei, Dongfang Yang, Licheng Wang, and Kai Wang. 2022. "A Method for Estimating the Aging State of Lithium-Ion Batteries Based on a Multi-Linear Integrated Model." *International Journal of Energy Research* 46 (15): 24091–24104. https://doi.org/10.1002/er.8709. <Go to ISI>://WOS:000853199600001.

Sun, Huiqian, Peng Jing, Baihui Wang, Yunhao Cai, Jie Ye, and Bichen Wang. 2023. "The Effect of Record-High Gasoline Prices on the Consumers? New Energy Vehicle Purchase Intention: Evidence from the Uniform Experimental Design." *Energy Policy* 175. https://doi.org/10.1016/j.enpol.2023.113500. <Go to ISI>://WOS:000946642800001.

Sun, Huiqin, Xiankui Wen, Wei Liu, Zhiqin Wang, and Qiangqiang Liao. 2022. “State-of-Health Estimation of Retired Lithium-Ion Battery Module Aged at 1C-Rate.” *Journal of Energy Storage* 50. https://doi.org/10.1016/j.est.2022.104618. <Go to ISI>://WOS:000793481500003.

Sun, Jiangtian, Xin Min, Jia-nan Gu, Jianxing Liang, and Mingming Guo. 2021. “Improved Performance of Mn-, Co-Based Oxides from Spent Lithium-Ion Batteries Supported on $CeO_2$ with Different Morphologies for 2-Ethoxyethyl Acetate Oxidation.” *Journal of Environmental Chemical Engineering* 9 (1). https://doi.org/10.1016/j.jece.2020.104964. <Go to ISI>://WOS:000615213100003.

Sun, Jinlei, Yong Tang, Jilei Ye, Tao Jiang, Saihan Chen, and Shengshi Qiu. 2022. “A Novel Capacity and Initial Discharge Electric Quantity Estimation Method for $LiFePO_4$ Battery Pack Based on OCV Curve Partial Reconstruction.” *Energy* 243. https://doi.org/10.1016/j.energy.2021.122882. <Go to ISI>://WOS:000791936900010.

Sun, Xinwei, Yang Zhang, Yongcheng Zhang, Licheng Wang, and Kai Wang. 2023. “Summary of Health-State Estimation of Lithium-Ion Batteries Based on Electrochemical Impedance Spectroscopy.” *Energies* 16 (15). https://doi.org/10.3390/en16155682. <Go to ISI>://WOS:001045398700001.

Sung, Woosuk, Do Sung Hwang, Jiwon Nam, Joo-Ho Choi, and Jaewook Lee. 2016. “Robust and Efficient Capacity Estimation Using Data-Driven Metamodel Applicable to Battery Management System of Electric Vehicles.” *Journal of the Electrochemical Society* 163 (6): A981–A991. https://doi.org/10.1149/2.0841606jes. <Go to ISI>://WOS:000373985300095.

Tabelin, Carlito Baltazar, Jessica Dallas, Sophia Casanova, Timothy Pelech, Ghislain Bournival, Serkan Saydam, and Ismet Canbulat. 2021. “Towards a Low-Carbon Society: A Review of Lithium Resource Availability, Challenges and Innovations in Mining, Extraction and Recycling, and Future Perspectives.” *Minerals Engineering* 163. https://doi.org/10.1016/j.mineng.2020.106743. <Go to ISI>://WOS:000756243000003.

Takyi-Aninakwa, Paul, Shunli Wang, Hongying Zhang, Yang Xiao, and Carlos Fernandez. 2023. “Enhanced Multi-State Estimation Methods for Lithium-Ion Batteries Considering Temperature Uncertainties.” *Journal of Energy Storage* 66. https://doi.org/10.1016/j.est.2023.107495. <Go to ISI>://WOS:000991589200001.

Taleb, Sylia Mekhmoukh, Yassine Meraihi, Seyedali Mirjalili, Dalila Acheli, Amar Ramdane-Cherif, and Asma Benmessaoud Gabis. 2023. “Mesh Router Nodes Placement for Wireless Mesh Networks Based on an Enhanced Moth-Flame Optimization Algorithm.” *Mobile Networks & Applications*. https://doi.org/10.1007/s11036-022-02059-6. <Go to ISI>://WOS:000909773000001.

Tan, Qingbo, Zhuning Wang, Wei Fan, Xudong Li, Xiangguang Li, Fanqi Li, and Zihao Zhao. 2023. “Development Path and Model Design of a New Energy Vehicle in China.” *Energies* 16 (1). https://doi.org/10.3390/en16010220. <Go to ISI>://WOS:000910297500001.

Tan, Shuang, Shangrui Zhao, and Jinran Wu. 2023. “QL-ADIFA: Hybrid Optimization Using Q-Learning and an Adaptive Logarithmic Spiral-Levy Firefly

Algorithm." *Mathematical Biosciences and Engineering* 20 (8): 13542–13561. https://doi.org/10.3934/mbe.2023604. <Go to ISI>://WOS:001017672800003.

Tang, Xiaopeng, Kailong Liu, Xin Wang, Furong Gao, James Macro, and W. Dhammika Widanage. 2020. "Model Migration Neural Network for Predicting Battery Aging Trajectories." *IEEE Transactions on Transportation Electrification* 6 (2): 363–374. https://doi.org/10.1109/tte.2020.2979547. <Go to ISI>://WOS:000545438200001.

Tawhid, Mohamed A., and Abdelmonem M. Ibrahim. 2022. "Improved Salp Swarm Algorithm Combined with Chaos." *Mathematics and Computers in Simulation* 202: 113–148. https://doi.org/10.1016/j.matcom.2022.05.029. <Go to ISI>://WOS:000822961000006.

Teng, Jen-Hao, Rong-Jhang Chen, Ping-Tse Lee, and Che-Wei Hsu. 2023. "Accurate and Efficient SOH Estimation for Retired Batteries." *Energies* 16 (3). https://doi.org/10.3390/en16031240. <Go to ISI>://WOS:000929522800001.

Tian, Huixin, Pengliang Qin, Kun Li, and Zhen Zhao. 2020. "A Review of the State of Health for Lithium-Ion Batteries: Research Status and Suggestions." *Journal of Cleaner Production* 261. https://doi.org/10.1016/j.jclepro.2020.120813. <Go to ISI>://WOS:000534478800004.

Tian, Jiaqiang, Siqi Li, Xinghua Liu, and Peng Wang. 2022. "Long-Short Term Memory Neural Network Based Life Prediction of Lithium-Ion Battery Considering Internal Parameters." *Energy Reports* 8: 81–89. https://doi.org/10.1016/j.egyr.2022.05.127. <Go to ISI>://WOS:000818068500009.

Tong, Xiaoxue, Tian Chen, Naiqiao Pan, and Xiangdong Zhang. 2023. "Quantum Combinational Logics and Their Realizations with Circuits." *Advanced Quantum Technologies*. https://doi.org/10.1002/qute.202300251. <Go to ISI>://WOS:001082671300001.

Tsai, Cheng-Tao, and Feng-Wei Peng. 2023. "Design and Implementation of Charging and Discharging Management System for Two-Set Lithium Ferrous Phosphate Batteries." *Sensors and Materials* 35 (4): 1255–1265. https://doi.org/10.18494/sam4119. <Go to ISI>://WOS:000966979100001.

Ungurean, Lucian, Mihai V. Micea, and Gabriel Carstoiu. 2020. "Online State of Health Prediction Method for Lithium-Ion Batteries, Based on Gated Recurrent Unit Neural Networks." *International Journal of Energy Research* 44 (8): 6767–6777. https://doi.org/10.1002/er.5413. <Go to ISI>://WOS:000524334200001.

Vahnstiege, Marc, Martin Winter, Sascha Nowak, and Simon Wiemers-Meyer. 2023. "State-of-Charge of Individual Active Material Particles in Lithium Ion Batteries: A Perspective of Analytical Techniques and Their Capabilities." *Physical Chemistry Chemical Physics*. https://doi.org/10.1039/d3cp02932. <Go to ISI>://WOS:001060310300001.

Van, Chi Nguyen, and Duy Ta Quang. 2023. "Estimation of SoH and Internal Resistances of Lithium Ion Battery Based on LSTM Network." *International Journal of Electrochemical Science* 18 (10). https://doi.org/10.1016/j.ijoes.2023.100166. <Go to ISI>://WOS:001033749300001.

Vitetta, Giorgio M., Pasquale Di Viesti, Emilio Sirignano, and Francesco Montorsi. 2020. "Multiple Bayesian Filtering as Message Passing." *IEEE*

*Transactions on Signal Processing* 68: 1002–1020. https://doi.org/10.1109/tsp.2020.2965296. <Go to ISI>://WOS:000526260900001.

Wang, Biao, Feifei Qin, Xiaobo Zhao, Xianpo Ni, and Dongji Xuan. 2020. "Equalization of Series Connected Lithium-Ion Batteries Based on Back Propagation Neural Network and Fuzzy Logic Control." *International Journal of Energy Research* 44 (6): 4812–4826. https://doi.org/10.1002/er.5274. <Go to ISI>://WOS:000517150700001.

Wang, Chao, Shunli Wang, Jinzhi Zhou, Jialu Qiao, Xiao Yang, and Yanxin Xie. 2023. "A Novel Back Propagation Neural Network-Dual Extended Kalman Filter Method for State-of-Charge and State-of-Health Co-Estimation of Lithium-Ion Batteries Based on Limited Memory Least Square Algorithm." *Journal of Energy Storage* 59. https://doi.org/10.1016/j.est.2022.106563. <Go to ISI>://WOS:000915861000001.

Wang, Cong-jie, Yan-li Zhu, Fei Gao, Chuang Qi, Peng-long Zhao, Qing-fen Meng, Jian-yong Wang, and Qi-bing Wu. 2020. "Thermal Runaway Behavior and Features of $LiFePO_4$/Graphite Aged Batteries under Overcharge." *International Journal of Energy Research* 44 (7): 5477–5487. https://doi.org/10.1002/er.5298. <Go to ISI>://WOS:000537950000027.

Wang, Genwei, Xuanfu Guo, Jingyi Chen, Pengfei Han, Qiliang Su, Meiqing Guo, Bin Wang, and Hui Song. 2023. "Safety Performance and Failure Criteria of Lithium-Ion Batteries under Mechanical Abuse." *Energies* 16 (17). https://doi.org/10.3390/en16176346. <Go to ISI>://WOS:001060541700001.

Wang, Guange, Huaning Zhang, Tong Wu, Borui Liu, Qing Huang, and Yuefeng Su. 2020. "Recycling and Regeneration of Spent Lithium-Ion Battery Cathode Materials." *Progress in Chemistry* 32 (12): 2064–2072. https://doi.org/10.7536/pc200119. <Go to ISI>://WOS:000608790500016.

Wang, Hao, Yanping Zheng, and Yang Yu. 2021. "Joint Estimation of SOC of Lithium Battery Based on Dual Kalman Filter." *Processes* 9 (8). https://doi.org/10.3390/pr9081412. <Go to ISI>://WOS:000689808000001.

Wang, Jiejia, Junrui Jia, and Zhihua Zhang. 2022. "Research on the Early Warning Mechanism for Thermal Runaway of Lithium-Ion Power Batteries in Electric Vehicles." *Security and Communication Networks* 2022. https://doi.org/10.1155/2022/8573396. <Go to ISI>://WOS:000812277200004.

Wang, Jingrong, Jinhao Meng, Qiao Peng, Tianqi Liu, Xueyang Zeng, Gang Chen, and Yan Li. 2023. "Lithium-Ion Battery State-of-Charge Estimation Using Electrochemical Model with Sensitive Parameters Adjustment." *Batteries-Basel* 9 (3). https://doi.org/10.3390/batteries9030180. <Go to ISI>://WOS:000954211500001.

Wang, Junlian, Jiashuai Fu, Fengshan Yu, Wen Xu, and Huajun Wang. 2020. "An Efficient Extractant (2-Ethylhexyl)(2,4,4′-Trimethylpentyl)Phosphinic Acid (USTB-1) for Cobalt and Nickel Separation from Sulfate Solutions." *Separation and Purification Technology* 248. https://doi.org/10.1016/j.seppur.2020.117060. <Go to ISI>://WOS:000538827600062.

Wang, Kai, Xiao Feng, Jinbo Pang, Jun Ren, Chongxiong Duan, and Liwei Li. 2020. "State of Charge (SOC) Estimation of Lithium-Ion Battery Based on Adaptive Square Root Unscented Kalman Filter." *International*

*Journal of Electrochemical Science* 15 (9): 9499–9516. https://doi.org/10.20964/2020.09.84. <Go to ISI>://WOS:000570963900088.

Wang, Nanlan, Xiangyang Xia, and Xiaoyong Zeng. 2023. "State of Charge and State of Health Estimation Strategies for Lithium-Ion Batteries." *International Journal of Low-Carbon Technologies* 18: 443–448. https://doi.org/10.1093/ijlct/ctad032. <Go to ISI>://WOS:000986675500001.

Wang, Qi, Tian Gao, and Xingcan Li. 2022. "SOC Estimation of Lithium-Ion Battery Based on Equivalent Circuit Model with Variable Parameters." *Energies* 15 (16). https://doi.org/10.3390/en15165829. <Go to ISI>://WOS:000846209100001.

Wang, Shunli, Jie Cao, Yanxin Xie, Haiying Gao, and Carlos Fernandez. 2022. "A Novel 2-RC Equivalent Model Based on the Self-Discharge Effect for Accurate State-of-Charge Estimation of Lithium-Ion Batteries." *International Journal of Electrochemical Science* 17 (7). https://doi.org/10.20964/2022.07.60. <Go to ISI>://WOS:000826841100018.

Wang, Shunli, Xianyi Jia, Paul Takyi-Aninakwa, Daniel-Ioan Stroe, and Carlos Fernandez. 2023. "Review-Optimized Particle Filtering Strategies for High-Accuracy State of Charge Estimation of LIBs." *Journal of the Electrochemical Society* 170 (5). https://doi.org/10.1149/1945-7111/acd148. <Go to ISI>://WOS:000985061500001.

Wang, Shunli, Paul Takyi-Aninakwa, Siyu Jin, Chunmei Yu, Carlos Fernandez, and Daniel-Ioan Stroe. 2022. "An Improved Feedforward-Long Short-Term Memory Modeling Method for the Whole-Life-Cycle State of Charge Prediction of Lithium-Ion Batteries Considering Current-Voltage-Temperature Variation." *Energy* 254. https://doi.org/10.1016/j.energy.2022.124224. <Go to ISI>://WOS:000807997900005.

Wang, Shunli, Paul Takyi-Aninakwa, Chunmei Yu, Siyu Jin, and Carlos Fernandez. 2022. "Improved Compound Correction-Electrical Equivalent Circuit Modeling and Double Transform-Unscented Kalman Filtering for the High-Accuracy Closed-Circuit Voltage and State-of-Charge Co-Estimation of Whole-Life-Cycle Lithium-Ion Batteries." *Energy Technology* 10 (12). https://doi.org/10.1002/ente.202200921. <Go to ISI>://WOS:000860084500001.

Wang, Shunli, Fan Wu, Paul Takyi-Aninakwa, Carlos Fernandez, Daniel-Ioan Stroe, and Qi Huang. 2023. "Improved Singular Filtering-Gaussian Process Regression-Long Short-Term Memory Model for Whole-Life-Cycle Remaining Capacity Estimation of Lithium-Ion Batteries Adaptive to Fast Aging and Multi-Current Variations." *Energy* 284: 1–16. https://doi.org/10.1016/j.energy.2023.128677. <Go to ISI>://WOS:001059157300001.

Wang, Xiaofei, and Qi Tong. 2023. "Simulating Fracture Patterns under Anisotropic Swelling in Lithiated Crystalline Nanostructures." *Engineering Fracture Mechanics* 281. https://doi.org/10.1016/j.engfracmech.2023.109088. <Go to ISI>://WOS:000963400600001.

Wang, Yuefei, Fei Huang, Bin Pan, Yang Li, and Baijun Liu. 2021. "Augmented System Model-Based Online Collaborative Determination of Lead-Acid Battery States for Energy Management of Vehicles." *Measurement & Control*

54 (1–2): 88–101. https://doi.org/10.1177/0020294020983376. <Go to ISI>://WOS:000614547400008.

Wang, Yujie, and Guanghui Zhao. 2023. "A Comparative Study of Fractional-Order Models for Lithium-Ion Batteries Using Runge Kutta Optimizer and Electrochemical Impedance Spectroscopy." *Control Engineering Practice* 133. https://doi.org/10.1016/j.conengprac.2023.105451. <Go to ISI>://WOS:000931819400001.

Wang, Zuolu, Guojin Feng, Dong Zhen, Fengshou Gu, and Andrew Ball. 2021. "A Review on Online State of Charge and State of Health Estimation for Lithium-Ion Batteries in Electric Vehicles." *Energy Reports* 7: 5141–5161. https://doi.org/10.1016/j.egyr.2021.08.113. <Go to ISI>://WOS:000701700500004.

Wang, Zuolu, Xiaoyu Zhao, Hao Zhang, Dong Zhen, Fengshou Gu, and Andrew Ball. 2023. "Active Acoustic Emission Sensing for Fast Co-Estimation of State of Charge and State of Health of the Lithium-Ion Battery." *Journal of Energy Storage* 64. https://doi.org/10.1016/j.est.2023.107192. <Go to ISI>://WOS:000961312200001.

Wang, Zuoxun, Xinheng Wang, Chunrui Ma, and Zengxu Song. 2021. "A Power Load Forecasting Model Based on FA-CSSA-ELM." *Mathematical Problems in Engineering* 2021. https://doi.org/10.1155/2021/9965932. <Go to ISI>://WOS:000664876500010.

Wei, Xiong, Yimin Mo, and Zhang Feng. 2019. "Lithium-Ion Battery Modeling and State of Charge Estimation." *Integrated Ferroelectrics* 200 (1): 59–72. https://doi.org/10.1080/10584587.2019.1592620. <Go to ISI>://WOS:000530565600007.

Wei, Zhen, Meng Li, Zhenchun Wei, Lei Cheng, Zengwei Lyu, and Fei Liu. 2020. "A Novel On-Demand Charging Strategy Based on Swarm Reinforcement Learning in WRSNs." *IEEE Access* 8: 84258–84271. https://doi.org/10.1109/access.2020.2992127. <Go to ISI>://WOS:000549526700003.

Willenberg, Lisa K., Philipp Dechent, Georg Fuchs, Dirk Uwe Sauer, and Egbert Figgemeier. 2020. "High-Precision Monitoring of Volume Change of Commercial Lithium-Ion Batteries by Using Strain Gauges." *Sustainability* 12 (2). https://doi.org/10.3390/su12020557. <Go to ISI>://WOS:000516824600115.

Wu, Che-Ya, Tzu-Ying Lin, and Jenq-Gong Duh. 2022. "Constructing Cellulose Nanofiber/Si Composite 3D-Net Structure from Waste Rice Straw Via Freeze Drying Process for Lithium-Ion Battery Anode Materials." *Materials Chemistry and Physics* 285. https://doi.org/10.1016/j.matchemphys.2022.126107. <Go to ISI>://WOS:000793571900001.

Wu, Fan, Shunli Wang, Wen Cao, Tao Long, Yawen Liang, and Carlos Fernandez. 2023. "An Improved Long Short-Term Memory Based on Global Optimization Square Root Extended Kalman Smoothing Algorithm for Collaborative State of Charge and State of Energy Estimation of Lithium-Ion Batteries." *International Journal of Circuit Theory and Applications* 51 (8): 3880–3896. https://doi.org/10.1002/cta.3624. <Go to ISI>://WOS:001042870500020.

Wu, Jie, Huigang Xu, and Peiyi Zhu. 2023. "State-of-Charge and State-of-Health Joint Estimation of Lithium-Ion Battery Based on Iterative Unscented

Kalman Particle Filtering Algorithm with Fused Rauch-Tung-Striebel Smoothing Structure." *Journal of Electrochemical Energy Conversion and Storage* 20 (4). https://doi.org/10.1115/1.4056557. <Go to ISI>://WOS:001078317000010.

Wu, Longxing, Kai Liu, Hui Pang, and Jiamin Jin. 2021. "Online SOC Estimation Based on Simplified Electrochemical Model for Lithium-Ion Batteries Considering Current Bias." *Energies* 14 (17): 5265–5277. https://doi.org/10.3390/en14175265. <Go to ISI>://WOS:000694069500001.

Wu, Muyao, Linlin Qin, and Gang Wu. 2021. "State of Charge Estimation of Power Lithium-Ion Battery Based on an Adaptive Time Scale Dual Extend Kalman Filtering." *Journal of Energy Storage* 39. https://doi.org/10.1016/j.est.2021.102535. <Go to ISI>://WOS:000659225900005.

Wu, Muyao, Li Wang, and Ji Wu. 2023. "State of Health Estimation of the LiFePO4 Power Battery Based on the Forgetting Factor Recursive Total Least Squares and the Temperature Correction." *Energy* 282. https://doi.org/10.1016/j.energy.2023.128437. <Go to ISI>://WOS:001042677800001.

Wu, Sheyin, Wenjie Pan, and Maotao Zhu. 2022. "A Collaborative Estimation Scheme for Lithium-Ion Battery State of Charge and State of Health Based on Electrochemical Model." *Journal of the Electrochemical Society* 169 (9). https://doi.org/10.1149/1945-7111/ac8ee4. <Go to ISI>://WOS:000854077400001.

Wu, Shujie, Rui Xiong, Hailong Li, Victor Nian, and Suxiao Ma. 2020. "The State of the Art on Preheating Lithium-Ion Batteries in Cold Weather." *Journal of Energy Storage* 27. https://doi.org/10.1016/j.est.2019.101059. <Go to ISI>://WOS:000516714200117.

Wu, Tiezhou, Sizhe Liu, Zhikun Wang, and Yiheng Huang. 2022. "SOC and SOH Joint Estimation of Lithium-Ion Battery Based on Improved Particle Filter Algorithm." *Journal of Electrical Engineering & Technology* 17 (1): 307–317. https://doi.org/10.1007/s42835-021-00861-y. <Go to ISI>://WOS:000683248700005.

Wu, Xiaoyu, Junjie Tang, Yuan Sun, and Yizhou Zhou. 2023. "Influence Mechanism of Phase Change on Leaching of Metal Elements from Ternary Lithium-Ion Battery Waste in Citric Acid." *Jom*. https://doi.org/10.1007/s11837-023-06083. <Go to ISI>://WOS:001063502100001.

Wu, Zhuoyan, Likun Yin, Ran Xiong, Shunli Wang, Wei Xiao, Yi Liu, Jun Jia, and Yanchao Liu. 2022. "A Novel State of Health Estimation of Lithium-Ion Battery Energy Storage System Based on Linear Decreasing Weight-Particle Swarm Optimization Algorithm and Incremental Capacity-Differential Voltage Method." *International Journal of Electrochemical Science* 17 (7). https://doi.org/10.20964/2022.07.41. <Go to ISI>://WOS:000826870800028.

Xia, Wei, Meiqiu Sun, and Zhuoyang Zhou. 2020. "Diffusion Collaborative Feedback Particle Filter." *IEEE Signal Processing Letters* 27: 1185–1189. https://doi.org/10.1109/lsp.2020.3003795. <Go to ISI>://WOS:000554896000001.

Xia, Zhiyong, and Jaber A. Abu Qahouq. 2021. "Lithium-Ion Battery Ageing Behavior Pattern Characterization and State-of-Health Estimation Using

Data-Driven Method." *IEEE Access* 9: 98287–98304. https://doi.org/10.1109/access.2021.3092743. <Go to ISI>://WOS:000673198300001.

Xiao, Dianxun, Gaoliang Fang, Sheng Liu, Shaoyi Yuan, Ryan Ahmed, Saeid Habibi, and Ali Emadi. 2020. "Reduced-Coupling Coestimation of SOC and SOH for Lithium-Ion Batteries Based on Convex Optimization." *IEEE Transactions on Power Electronics* 35 (11): 12332–12346. https://doi.org/10.1109/tpel.2020.2984248. <Go to ISI>://WOS:000555006800087.

Xiao, Renxin, Yanwen Hu, Wei Zhang, and Zhaohui Chen. 2023. "A Novel Approach to Estimate the State of Charge for Lithium-Ion Battery under Different Temperatures Incorporating Open Circuit Voltage Online Identification." *Journal of Energy Storage* 67. https://doi.org/10.1016/j.est.2023.107509. <Go to ISI>://WOS:001001069600001.

Xie, Baoshan, Fei Li, Hao Li, Liya Wang, and Aimin Yang. 2023. "Enhanced Internet of Things Security Situation Assessment Model with Feature Optimization and Improved SSA-LightGBM." *Mathematics* 11 (16). https://doi.org/10.3390/math11163617. <Go to ISI>://WOS:001055228800001.

Xie, Jiamiao, Xingyu Wei, Xiqiao Bo, Peng Zhang, Pengyun Chen, Wenqian Hao, and Meini Yuan. 2023. "State of Charge Estimation of Lithium-Ion Battery Based on Extended Kalman Filter Algorithm." *Frontiers in Energy Research* 11. https://doi.org/10.3389/fenrg.2023.1180881. <Go to ISI>://WOS:001000387600001.

Xie, Yanxin, Shunli Wang, Carlos Fernandez, Chunmei Yu, Yongcun Fan, and Wen Cao. 2021. "A Novel High-Fidelity Unscented Particle Filtering Method for the Accurate State of Charge Estimation of Lithium-Ion Batteries." *International Journal of Electrochemical Science* 16 (6). https://doi.org/10.20964/2021.06.38. <Go to ISI>://WOS:000661490000023.

Xie, Yanxin, Shunli Wang, Gexiang Zhang, Yongcun Fan, Carlos Fernandez, and Frede Blaabjerg. 2023. "Optimized Multi-Hidden Layer Long Short-Term Memory Modeling and Suboptimal Fading Extended Kalman Filtering Strategies for the Synthetic State of Charge Estimation of Lithium-Ion Batteries." *Applied Energy* 336. https://doi.org/10.1016/j.apenergy.2023.120866. <Go to ISI>://WOS:000996192300001.

Xie, Yi, Wei Li, Xiaosong Hu, Manh-Kien Tran, Satyam Panchal, Michael Fowler, Yangjun Zhang, and Kailong Liu. 2023. "Coestimation of SOC and Three-Dimensional SOT for Lithium-Ion Batteries Based on Distributed Spatial-Temporal Online Correction." *IEEE Transactions on Industrial Electronics* 70 (6): 5937–5948. https://doi.org/10.1109/tie.2022.3199905. <Go to ISI>://WOS:000966134500001.

Xie, Yi, Xi Wang, Xiaosong Hu, Wei Li, Yangjun Zhang, and Xianke Lin. 2022. "An Enhanced Electro-Thermal Model for EV Battery Packs Considering Current Distribution in Parallel Branches." *IEEE Transactions on Power Electronics* 37 (1): 1027–1043. https://doi.org/10.1109/tpel.2021.3102292. <Go to ISI>://WOS:000698580400093.

Xiong, Rui, Binyu Xiong, Qingyong Zhang, Shaoyue Shi, Yixin Su, and Danhong Zhang. 2022. "Capacity Fading Model of Vanadium Redox Flow Battery Considering Water Molecules Migration." *International Journal of Green*

*Energy* 19 (15): 1613–1622. https://doi.org/10.1080/15435075.2021.2015599. <Go to ISI>://WOS:000745413600001.

Xu, Cheng, E. Zhang, Kai Jiang, and Kangli Wang. 2022. "Dual Fuzzy-Based Adaptive Extended Kalman Filter for State of Charge Estimation of Liquid Metal Battery." *Applied Energy* 327. https://doi.org/10.1016/j.apenergy.2022.120091. <Go to ISI>://WOS:000930578100004.

Xu, Chunyu, Huipeng Meng, and Yufeng Wang. 2020. "A Novel Hybrid Firefly Algorithm Based on the Vector Angle Learning Mechanism." *IEEE Access* 8: 205741–205754. https://doi.org/10.1109/access.2020.3037802. <Go to ISI>://WOS:000594425500001.

Xu, Guanghui, Ting-Wei Zhang, Qiang Lai, Jian Pan, Bo Fu, and Xilin Zhao. 2020. "A New Path Planning Method of Mobile Robot Based on Adaptive Dynamic Firefly Algorithm." *Modern Physics Letters B* 34 (29). https://doi.org/10.1142/s0217984920503224. <Go to ISI>://WOS:000585949700006.

Xu, Jianguang, Menglan Jin, Xinlu Shi, Qiuyu Li, Chengqiang Gan, and Wei Yao. 2021. "Preparation of $TiSi_2$ Powders with Enhanced Lithium-Ion Storage via Chemical Oven Self-Propagating High-Temperature Synthesis." *Nanomaterials* 11 (9). https://doi.org/10.3390/nano11092279. <Go to ISI>://WOS:000701410900001.

Xu, Jieyu, and Dongqing Wang. 2022. "A Dual-Rate Sampled Multiple Innovation Adaptive Extended Kalman Filter Algorithm for State of Charge Estimation." *International Journal of Energy Research* 46 (13): 18796–18808. https://doi.org/10.1002/er.8498. <Go to ISI>://WOS:000851534500001.

Xu, Jingjing, Xingyun Cai, Songming Cai, Yaxin Shao, Chao Hu, Shirong Lu, and Shujiang Ding. 2023. "High-Energy Lithium-Ion Batteries: Recent Progress and a Promising Future in Applications." *Energy & Environmental Materials* 6 (5). https://doi.org/10.1002/eem2.12450. <Go to ISI>://WOS:000913157100001.

Xu, Jinmei, Shengkun Xie, Zhen Lin, Xiangyun Qiu, Kai Wu, and Honghe Zheng. 2022. "Studies of Interfacial Reaction Characteristics for High Power Lithium-Ion Battery." *Electrochimica Acta* 435. https://doi.org/10.1016/j.electacta.2022.141305. <Go to ISI>://WOS:000882440100002.

Xu, Jinmei, Jiandong Yang, Shaofei Wang, Jiangmin Jiang, Quanchao Zhuang, Xiangyun Qiu, Kai Wu, and Honghe Zheng. 2023. "Ion Transport and Electrochemical Reaction in $LiNi_{0.5}Co_{0.2}Mn_{0.3}O_2$-Based High Energy/Power Lithium-Ion Batteries." *Nanomaterials* 13 (5). https://doi.org/10.3390/nano13050856. <Go to ISI>://WOS:000947413500001.

Xu, Jun, Xuesong Mei, Xiao Wang, Yumeng Fu, Yunfei Zhao, and Junping Wang. 2021. "A Relative State of Health Estimation Method Based on Wavelet Analysis for Lithium-Ion Battery Cells." *IEEE Transactions on Industrial Electronics* 68 (8): 6973–6981. https://doi.org/10.1109/tie.2020.3001836. <Go to ISI>://WOS:000647484000054.

Xu, Liuchao, Zhiheng Pan, Chuandong Liang, and Min Lu. 2022. "A Fault Diagnosis Method for PV Arrays Based on New Feature Extraction and Improved the Fuzzy C-Mean Clustering." *IEEE Journal of Photovoltaics* 12 (3): 833–843. https://doi.org/10.1109/jphotov.2022.3151330. <Go to ISI>://WOS:000764851600001.

Xu, Ruilong, Yujie Wang, and Zonghai Chen. 2022. "A Migration-Based Method for Non-Invasive Revelation of Microscopic Degradation Mechanisms and Health Prognosis of Lithium-Ion Batteries." *Journal of Energy Storage* 55. https://doi.org/10.1016/j.est.2022.105769. <Go to ISI>://WOS:000872546800004.

Xu, Wenhua, Shunli Wang, Carlos Fernandez, Chunmei Yu, Yongcun Fan, and Wen Cao. 2020. "Novel Reduced-Order Modeling Method Combined with Three-Particle Nonlinear Transform Unscented Kalman Filtering for the Battery State-of-Charge Estimation." *Journal of Power Electronics* 20 (6): 1541–1549. https://doi.org/10.1007/s43236-020-00146-z. <Go to ISI>://WOS:000570511800003.

Xu, Yonghong, Xia Chen, Hongguang Zhang, Fubin Yang, Liang Tong, Yifan Yang, Dong Yan, Anren Yang, Mingzhe Yu, Zhuxian Liu, and Yan Wang. 2022. "Online Identification of Battery Model Parameters and Joint State of Charge and State of Health Estimation Using Dual Particle Filter Algorithms." *International Journal of Energy Research* 46 (14): 19615–19652. https://doi.org/10.1002/er.8541. <Go to ISI>://WOS:000842176000001.

Xu, Yue, Yang Li, Yong Qian, Shanshan Sun, Ning Lin, and Yitai Qian. 2023. "Deficient $TiO_2.x$ Coated Porous SiO Anodes for High-Rate Lithium-Ion Batteries." *Inorganic Chemistry Frontiers* 10 (4): 1176–1186. https://doi.org/10.1039/d2qi02447k. <Go to ISI>://WOS:000914549000001.

Xu, Zhicheng, Jun Wang, Peter D. Lund, and Yaoming Zhang. 2022. "Co-Estimating the State of Charge and Health of Lithium Batteries through Combining a Minimalist Electrochemical Model and an Equivalent Circuit Model." *Energy* 240. https://doi.org/10.1016/j.energy.2021.122815. <Go to ISI>://WOS:000738791300011.

Yan, Xuena, Shunfu Jin, Wuyi Yue, and Yutaka Takahashi. 2021. "Performance Analysis and System Optimization of an Energy-Saving Mechanism in Cloud Computing with Correlated Traffic." *Journal of Industrial and Management Optimization*. https://doi.org/10.3934/jimo.2021106. <Go to ISI>://WOS:000706701300001.

Yang, Cheng, Jialiang Zhang, Guoqiang Liang, Hao Jin, Yongqiang Chen, and Chengyan Wang. 2022. "An Advanced Strategy of? Metallurgy before Sorting? For Recycling Spent Entire Ternary Lithium-Ion Batteries." *Journal of Cleaner Production* 361. https://doi.org/10.1016/j.jclepro.2022.132268. <Go to ISI>://WOS:000807777600003.

Yang, Fan, Dongliang Shi, Qian Mao, and Kwok-ho Lam. 2023. "Scientometric Research and Critical Analysis of Battery State-of-Charge Estimation." *Journal of Energy Storage* 58. https://doi.org/10.1016/j.est.2022.106283. <Go to ISI>://WOS:000909175100001.

Yang, Jing, Bin Liu, and Lele Duan. 2020. "Structural Evolution of the Ru-Bms Complex to the Real Water Oxidation Catalyst of Ru-Bda: The Bite Angle Matters." *Dalton Transactions* 49 (14): 4369–4375. https://doi.org/10.1039/c9dt04693c. <Go to ISI>://WOS:000526110700019.

Yang, Niankai, Ziyou Song, Heath Hofmann, and Jing Sun. 2022. "Robust State of Health Estimation of Lithium-Ion Batteries Using Convolutional Neural

Network and Random Forest." *Journal of Energy Storage* 48. https://doi.org/10.1016/j.est.2021.103857. <Go to ISI>://WOS:000780272100002.

Yang, Sijia, Caiping Zhang, Jiuchun Jiang, Weige Zhang, Linjing Zhang, and Yubin Wang. 2021. "Review on State-of-Health of Lithium-Ion Batteries: Characterizations, Estimations and Applications." *Journal of Cleaner Production* 314. https://doi.org/10.1016/j.jclepro.2021.128015. <Go to ISI>://WOS:000688461400004.

Yang, Xiaoyong, Shunli Wang, Paul Takyi-Aninakwa, Xiao Yang, and Carlos Fernandez. 2023. "Improved Noise Bias Compensation-Equivalent Circuit Modeling Strategy for Battery State of Charge Estimation Adaptive to Strong Electromagnetic Interference." *Journal of Energy Storage* 73. https://doi.org/10.1016/j.est.2023.108974. <Go to ISI>://WOS:001079479900001.

Yang, Yalong, Siyuan Chen, Tao Chen, and Liansheng Huang. 2023. "State of Health Assessment of Lithium-Ion Batteries Based on Deep Gaussian Process Regression Considering Heterogeneous Features." *Journal of Energy Storage* 61. https://doi.org/10.1016/j.est.2023.106797. <Go to ISI>://WOS:001008923000001.

Yang, Yang, Libo Lan, Zhuo Hao, Jianyou Zhao, Geng Luo, Pei Fu, and Yisong Chen. 2022. "Life Cycle Prediction Assessment of Battery Electrical Vehicles with Special Focus on Different Lithium-Ion Power Batteries in China." *Energies* 15 (15). https://doi.org/10.3390/en15155321. <Go to ISI>://WOS:000839913300001.

Yang, Yun, Zhirong Wang, Juncheng Jiang, Huan Bian, Ning Mao, and Linsheng Guo. 2020. "Effects of Different Charging and Discharging Modes on Thermal Behavior of Lithium Ion Batteries." *Fire and Materials* 44 (1): 90–99. https://doi.org/10.1002/fam.2778. <Go to ISI>://WOS:000497479000001.

Yao, Hang, Xiang Jia, Qian Zhao, Zhi-Jun Cheng, and Bo Guo. 2020. "Novel Lithium-Ion Battery State-of-Health Estimation Method Using a Genetic Programming Model." *IEEE Access* 8: 95333–95344. https://doi.org/10.1109/access.2020.2995899. <Go to ISI>://WOS:000541139500013.

Yao, Lei, Jishu Wen, Shiming Xu, Jie Zheng, Junjian Hou, Zhanpeng Fang, and Yanqiu Xiao. 2022. "State of Health Estimation Based on the Long Short-Term Memory Network Using Incremental Capacity and Transfer Learning." *Sensors* 22 (20). https://doi.org/10.3390/s22207835. <Go to ISI>://WOS:000873872500001.

Ye, Lihua, Dinghan Peng, Dingbang Xue, Sijian Chen, and Aiping Shi. 2023. "Co-Estimation of Lithium-Ion Battery State-of-Charge and State-of-Health Based on Fractional-Order Model." *Journal of Energy Storage* 65. https://doi.org/10.1016/j.est.2023.107225. <Go to ISI>://WOS:000980734600001.

Ye, Yiming, Jiangfeng Zhang, Srikanth Pilla, Apparao M. Rao, and Bin Xu. 2023. "Application of a New Type of Lithium-Sulfur Battery and Reinforcement Learning in Plug-in Hybrid Electric Vehicle Energy Management." *Journal of Energy Storage* 59. https://doi.org/10.1016/j.est.2022.106546. <Go to ISI>://WOS:000915769000001.

Yeong, Hoong C., Ryne T. Beeson, N. Sri Namachchivaya, and Nicolas Perkowski. 2020. "Particle Filters with Nudging in Multiscale Chaotic Systems: With

Application to the Lorenz '96 Atmospheric Model." *Journal of Nonlinear Science* 30 (4): 1519–1552. https://doi.org/10.1007/s00332-020-09616-x. <Go to ISI>://WOS:000516352500001.

Yi, Lingzhi, Ganlin Jiang, Guoyong Zhang, Wenxin Yu, You Guo, and Tao Sun. 2022. "A Fault Diagnosis Method of Oil-Immersed Transformer Based on Improved Harris Hawks Optimized Random Forest." *Journal of Electrical Engineering & Technology* 17 (4): 2527–2540. https://doi.org/10.1007/s42835-022-01036-z. <Go to ISI>://WOS:000771895000004.

Yin, Tao, Longzhou Jia, Xichao Li, Lili Zheng, and Zuoqiang Dai. 2022. "Effect of High-Rate Cycle Aging and Over-Discharge on NCM811 (LiNi0.8Co0.1Mn0.1O2) Batteries." *Energies* 15 (8). https://doi.org/10.3390/en15082862. <Go to ISI>://WOS:000786804900001.

Yoneda, Tetsuya. 2023. "Development of a Polychromatic Simultaneous Wavelength Dispersive X-Ray Spectrometer and Improvement in Its Precision: From Principle Verification to Chemical State Analysis of Cathode Materials in Lithium-Ion Batteries." *Bunseki Kagaku* 72 (1–2): 33–43. <Go to ISI>://WOS:000926481100001.

Yong-Taek, Han, Il-Won Kim, and Kim Si-Kuk. 2022. "A Study on the Fire Risk Due to Overcharging of Self-Made Lithium-Ion Battery Packs." *Korean Journal of Hazardous Materials* 10 (1): 73–81. <Go to ISI>://KJD:ART002856147.

Yoshimura, Aya, Keisuke Hemmi, Hayato Moriwaki, Ryo Sakakibara, Hitoshi Kimura, Yuto Aso, Naoya Kinoshita, Rie Suizu, Takashi Shirahata, Masaru Yao, Hideki Yorimitsu, Kunio Awaga, and Yohji Misaki. 2022. "Improvement in Cycle Life of Organic Lithium-Ion Batteries by In-Cell Polymerization of Tetrathiafulvalene-Based Electrode Materials." *ACS Applied Materials & Interfaces* 14 (31): 35978–35984. https://doi.org/10.1021/acsami.2c09302. <Go to ISI>://WOS:000834273600001.

You, Heze, Haifeng Dai, Lizhen Li, Xuezhe Wei, and Guangshuai Han. 2021. "Charging Strategy Optimization at Low Temperatures for Li-Ion Batteries Based on Multi-Factor Coupling Aging Model." *IEEE Transactions on Vehicular Technology* 70 (11): 11433–11445. https://doi.org/10.1109/tvt.2021.3114298. <Go to ISI>://WOS:000720520400027.

Yu, Chuanxiang, Rui Huang, Zhaoyu Sang, and Shiya Yang. 2022. "A Novel Trigger Mechanism for a Dual-Filter to Improve the State-of-Charge Estimation of Lithium-Ion Batteries." *Journal of Electrochemical Energy Conversion and Storage* 19 (3). https://doi.org/10.1115/1.4052993. <Go to ISI>://WOS:000818094900017.

Yu, Da, Dongsheng Ren, Keren Dai, He Zhang, Jinming Zhang, Benqiang Yang, Shaojie Ma, Xiaofeng Wang, and Zheng You. 2021. "Failure Mechanism and Predictive Model of Lithium-Ion Batteries under Extremely High Transient Impact." *Journal of Energy Storage* 43. https://doi.org/10.1016/j.est.2021.103191. <Go to ISI>://WOS:000701654600003.

Yu, Jiuling. 2020. *Metal Oxide and Carbonaceous Material for Lithium Ion Battery.* Edited by Jiuling Yu. Ann Arbor: ProQuest Dissertations Publishing.

Yu, Junqi, Qite Liu, Anjun Zhao, Xuegen Qian, and Rui Zhang. 2020. "Optimal Chiller Loading in HVAC System Using a Novel Algorithm Based on the

Distributed Framework." *Journal of Building Engineering* 28. https://doi.org/10.1016/j.jobe.2019.101044. <Go to ISI>://WOS:000513823200063.

Yu, Shuhao, Xukun Zuo, Xianglin Fan, Zhengyu Liu, and Mingjing Pei. 2021. "An Improved Firefly Algorithm Based on Personalized Step Strategy." *Computing* 103 (4): 735–748. https://doi.org/10.1007/s00607-021-00919-9. <Go to ISI>://WOS:000620849000001.

Yuan, Baohe, Binger Zhang, Xiang Yuan, Jingyi Wang, Lulu Chen, Lei Bai, and Shijun Luo. 2022. "Study on the Relationship between Open-Circuit Voltage, Time Constant and Polarization Resistance of Lithium-Ion Batteries." *Journal of the Electrochemical Society* 169 (6). https://doi.org/10.1149/1945-7111/ac7359. <Go to ISI>://WOS:000806528800001.

Yuan, Qiuqi, Xiaoming Xu, Lei Zhao, Guangyao Tong, and Lei Zhu. 2020. "Multitime Scale Analysis of Surface Temperature Distribution of Lithium-Ion Batteries in Quantity-Quality Change under Local High-Temperature Heat Source." *Journal of Energy Engineering* 146 (6). https://doi.org/10.1061/(asce)ey.1943-7897.0000706. <Go to ISI>://WOS:000609456500004.

Yue, Q. L., C. X. He, M. C. Wu, and T. S. Zhao. 2021. "Advances in Thermal Management Systems for Next-Generation Power Batteries." *International Journal of Heat and Mass Transfer* 181. https://doi.org/10.1016/j.ijheatmasstransfer.2021.121853. <Go to ISI>://WOS:000706121000005.

Yun, Seok-Teak, and Seung-Hyun Kong. 2022. "Data-Driven in-Orbit Current and Voltage Prediction Using Bi-LSTM for LEO Satellite Lithium-Ion Battery SOC Estimation." *IEEE Transactions on Aerospace and Electronic Systems* 58 (6): 5292–5306. https://doi.org/10.1109/taes.2022.3167624. <Go to ISI>://WOS:000895081000031.

Zeng, Taotao, Dai-Huo Liu, Changling Fan, Runzheng Fan, Fuquan Zhang, Jinshui Liu, Tingzhou Yang, and Zhongwei Chen. 2023. "$LiMn_{0.8}Fe_{0.2}PO_4$@C Cathode Prepared *via* a Novel Hydrated $MnHPO_4$ Intermediate for High Performance Lithium-Ion Batteries." *Inorganic Chemistry Frontiers* 10 (4): 1164–1175. https://doi.org/10.1039/d2qi02306g. <Go to ISI>://WOS:000916536900001.

Zhang, Chaolong, Shaishai Zhao, Zhong Yang, and Yuan Chen. 2022. "A Reliable Data-Driven State-of-Health Estimation Model for Lithium-Ion Batteries in Electric Vehicles." *Frontiers in Energy Research* 10. https://doi.org/10.3389/fenrg.2022.1013800. <Go to ISI>://WOS:000868659000001.

Zhang, Dan, Chao Tan, Ting Ou, Shengrui Zhang, Le Li, and Xiaohui Ji. 2022. "Constructing Advanced Electrode Materials for Low-Temperature Lithium-Ion Batteries: A Review." *Energy Reports* 8: 4525–4534. https://doi.org/10.1016/j.egyr.2022.03.130. <Go to ISI>://WOS:000792926900008.

Zhang, Guowei, Zheng Li, Hongyu Wang, and Diping Yuan. 2022. "Study on the Suppression Effect of Cryogenic Cooling on Thermal Runaway of Ternary Lithium-Ion Batteries." *Fire-Switzerland* 5 (6). https://doi.org/10.3390/fire5060182. <Go to ISI>://WOS:000901296100001.

Zhang, Hao, and Guixin Cai. 2020. "Subsidy Strategy on New-Energy Vehicle Based on Incomplete Information: A Case in China." *Physica a-Statistical Mechanics and Its Applications* 541. https://doi.org/10.1016/j.physa.2019.123370. <Go to ISI>://WOS:000514758600041.

Zhang, Haoyi, Fuquan Zhao, Han Hao, and Zongwei Liu. 2021. "Effect of Chinese Corporate Average Fuel Consumption and New Energy Vehicle Dual-Credit Regulation on Passenger Cars Average Fuel Consumption Analysis." *International Journal of Environmental Research and Public Health* 18 (14). https://doi.org/10.3390/ijerph18147218. <Go to ISI>://WOS:000677350300001.

Zhang, Jian Hua, Yu Hong Jin, Jing Bing Liu, Qian Qian Zhang, and Hao Wang. 2021. "Recent Advances in Understanding and Relieving Capacity Decay of Lithium Ion Batteries with Layered Ternary Cathodes." *Sustainable Energy & Fuels* 5 (20): 5114–5138. https://doi.org/10.1039/d1se01137e. <Go to ISI>://WOS:000700555100001.

Zhang, Lupeng, Xinle Li, Mingrui Yang, and Weihua Chen. 2021. "High-Safety Separators for Lithium-Ion Batteries and Sodium-Ion Batteries: Advances and Perspective." *Energy Storage Materials* 41: 522–545. https://doi.org/10.1016/j.ensm.2021.06.033. <Go to ISI>://WOS:000685110200006.

Zhang, Mengyun, Shunli Wang, Xiao Yang, Yanxin Xie, Ke Liu, and Chuyan Zhang. 2023. "Improved Backward Smoothing-Square Root Cubature Kalman Filtering and Variable Forgetting Factor-Recursive Least Square Modeling Methods for the High-Precision State of Charge Estimation of Lithium-Ion Batteries." *Journal of the Electrochemical Society* 170 (3). https://doi.org/10.1149/1945-7111/acb10b. <Go to ISI>://WOS:000952062800001.

Zhang, Qunming, Cheng-Geng Huang, He Li, Guodong Feng, and Weiwen Peng. 2022. "Electrochemical Impedance Spectroscopy Based State-of-Health Estimation for Lithium-Ion Battery Considering Temperature and State-of-Charge Effect." *IEEE Transactions on Transportation Electrification* 8 (4): 4633–4645. https://doi.org/10.1109/tte.2022.3160021. <Go to ISI>://WOS:000871082600050.

Zhang, Shuzhi, Nian Peng, Haibin Lu, Rui Li, and Xiongwen Zhang. 2022. "A Systematic and Low-Complexity Multi-State Estimation Framework for Series-Connected Lithium-Ion Battery Pack under Passive Balance Control." *Journal of Energy Storage* 48. https://doi.org/10.1016/j.est.2022.103989. <Go to ISI>://WOS:000780357800004.

Zhang, Tao, Ningyuan Guo, Xiaoxia Sun, Jie Fan, Naifeng Yang, Junjie Song, and Yuan Zou. 2021. "A Systematic Framework for State of Charge, State of Health and State of Power Co-Estimation of Lithium-Ion Battery in Electric Vehicles." *Sustainability* 13 (9). https://doi.org/10.3390/su13095166. <Go to ISI>://WOS:000650915200001.

Zhang, Wencan, Taotao Li, Weixiong Wu, Nan Ouyang, and Guangshan Huang. 2023. "Data-Driven State of Health Estimation in Retired Battery Based on Low and Medium-Frequency Electrochemical Impedance Spectroscopy." *Measurement* 211. https://doi.org/10.1016/j.measurement.2023.112597. <Go to ISI>://WOS:000964621500001.

Zhang, Xiaodong, Jing Sun, Yunlong Shang, Song Ren, Yiwei Liu, and Diantao Wang. 2022. "A Novel State-of-Health Prediction Method Based on Long Short-Term Memory Network with Attention Mechanism for Lithium-Ion Battery." *Frontiers in Energy Research* 10. https://doi.org/10.3389/fenrg.2022.972486. <Go to ISI>://WOS:000853987700001.

Zhang, Xiaoqiang, and Gongxing Yan. 2021. "Estimating SOC and SOH of Lithium Battery Based on Nano Material." *Ferroelectrics* 580 (1): 112–128. https://doi.org/10.1080/00150193.2021.1905731. <Go to ISI>://WOS:000695478400009.

Zhang, Xing, Zhibin Yan, and Yunqi Chen. 2022. "High-Degree Cubature Kalman Filter for Nonlinear State Estimation with Missing Measurements." *Asian Journal of Control* 24 (3): 1261–1272. https://doi.org/10.1002/asjc.2510. <Go to ISI>://WOS:000609091100001.

Zhang, Yajun, Yajie Liu, Jia Wang, and Tao Zhang. 2022. "State-of-Health Estimation for Lithium-Ion Batteries by Combining Model-Based Incremental Capacity Analysis with Support Vector Regression." *Energy* 239. https://doi.org/10.1016/j.energy.2021.121986. <Go to ISI>://WOS:000706253800004.

Zhang, Zhengjie, Yefan Sun, Lisheng Zhang, Hanchao Cheng, Rui Cao, Xinhua Liu, and Shichun Yang. 2023. "Enabling Online Search and Fault Inference for Batteries Based on Knowledge Graph." *Batteries-Basel* 9 (2). https://doi.org/10.3390/batteries9020124. <Go to ISI>://WOS:000938219000001.

Zhao, Bin, Hao Chen, Diankui Gao, and Lizhi Xu. 2020. "Risk Assessment of Refinery Unit Maintenance Based on Fuzzy Second Generation Curvelet Neural Network." *Alexandria Engineering Journal* 59 (3): 1823–1831. https://doi.org/10.1016/j.aej.2020.04.052. <Go to ISI>://WOS:000542589300007.

Zhao, Guanghui, Yujie Wang, and Zonghai Chen. 2022. "Health-Aware Multi-Stage Charging Strategy for Lithium-Ion Batteries Based on Whale Optimization Algorithm." *Journal of Energy Storage* 55. https://doi.org/10.1016/j.est.2022.105620. <Go to ISI>://WOS:000869785600002.

Zhao, Guoqi, Yu Liu, Gang Liu, Shiping Jiang, and Wenfeng Hao. 2021. "State-of-Charge and State-of-Health Estimation for Lithium-Ion Battery Using the Direct Wave Signals of Guided Wave." *Journal of Energy Storage* 39. https://doi.org/10.1016/j.est.2021.102657. <Go to ISI>://WOS:000663786500003.

Zhao, Jiahui, Yong Zhu, Bin Zhang, Mingyi Liu, Jianxing Wang, Chenghao Liu, and Xiaowei Hao. 2023. "Review of State Estimation and Remaining Useful Life Prediction Methods for Lithium-Ion Batteries." *Sustainability* 15 (6). https://doi.org/10.3390/su15065014. <Go to ISI>://WOS:000959858600001.

Zhao, Qing, Lv Hu, Wenjie Li, Chengjun Liu, Maofa Jiang, and Junjie Shi. 2020. "Recovery and Regeneration of Spent Lithium-Ion Batteries from New Energy Vehicles." *Frontiers in Chemistry* 8. https://doi.org/10.3389/fchem.2020.00807. <Go to ISI>://WOS:000588486400001.

Zhao, Runmin, Jinzhi Gong, Yangzezhi Zheng, and Xiaoming Huang. 2022. "Research on the Combination of Firefly Intelligent Algorithm and Asphalt Material Modulus Back Calculation." *Materials* 15 (9). https://doi.org/10.3390/ma15093361. <Go to ISI>://WOS:000799264600001.

Zhao, Xiaobo, Seunghun Jung, Biao Wang, and Dongji Xuan. 2023. "State of Charge Estimation of Lithium-Ion Battery Based on Improved Adaptive Boosting Algorithm." *Journal of Energy Storage* 71. https://doi.org/10.1016/j.est.2023.108047. <Go to ISI>://WOS:001053841300001.

Zhao, Xu, Jianyao Hu, Guangdi Hu, and Huimin Qiu. 2023. "A State of Health Estimation Framework Based on Real-World Electric Vehicles Operating

Data." *Journal of Energy Storage* 63. https://doi.org/10.1016/j.est.2023.107031. <Go to ISI>://WOS:000952260300001.

Zhi, Maoyong, Rong Fan, Xiong Yang, Lingling Zheng, Shan Yue, Quanyi Liu, and Yuanhua He. 2022. "Recent Research Progress on Phase Change Materials for Thermal Management of Lithium-Ion Batteries." *Journal of Energy Storage* 45. https://doi.org/10.1016/j.est.2021.103694. <Go to ISI>://WOS:000735336300001.

Zhou, Xinwei, Junqi Yu, Wanhu Zhang, Anjun Zhao, and Min Zhou. 2022. "A Multi-Objective Optimization Operation Strategy for Ice-Storage Air-Conditioning System Based on Improved Firefly Algorithm." *Building Services Engineering Research & Technology* 43 (2): 161–178. https://doi.org/10.1177/01436244211045570. <Go to ISI>://WOS:000708031300001.

Zhou, Yu, Hua Deng, and Han-Xiong Li. 2023. "Control-Oriented Galerkin-Spectral Model for 3-D Thermal Diffusion of Pouch-Type Batteries." *IEEE Transactions on Industrial Informatics* 19 (6): 7508–7516. https://doi.org/10.1109/tii.2022.3212279. <Go to ISI>://WOS:001054612400017.

Zhou, Zhaihe, Yulu Zhong, Chuanwei Zeng, and Xiangrui Tian. 2021. "Attitude Estimation Using Parallel Quaternion Particle Filter Based on New Quaternion Distribution." *Transactions of the Japan Society for Aeronautical and Space Sciences* 64 (5): 249–257. https://doi.org/10.2322/tjsass.64.249. <Go to ISI>://WOS:000692037000001.

Zhu, Chenyu, Shunli Wang, Chunmei Yu, Heng Zhou, and Carlos Fernandez. 2023. "An Improved Proportional Control Forgetting Factor Recursive Least Square-Monte Carlo Adaptive Extended Kalman Filtering Algorithm for High-Precision State-of-Charge Estimation of Lithium-Ion Batteries." *Journal of Solid State Electrochemistry* 27 (9): 2277–2287. https://doi.org/10.1007/s10008-023-05514-w. <Go to ISI>://WOS:000980350400002.

Zhu, Feng, and Jingqi Fu. 2021. "A Novel State-of-Health Estimation for Lithium-Ion Battery via Unscented Kalman Filter and Improved Unscented Particle Filter." *IEEE Sensors Journal* 21 (22): 25449–25456. https://doi.org/10.1109/jsen.2021.3102990. <Go to ISI>://WOS:000717802500065.

Zhu, Tao, Shunli Wang, Yongcun Fan, Heng Zhou, Yifei Zhou, and Carlos Fernandez. 2023. "Improved Forgetting Factor Recursive Least Square and Adaptive Square Root Unscented Kalman Filtering Methods for Online Model Parameter Identification and Joint Estimation of State of Charge and State of Energy of Lithium-Ion Batteries." *Ionics*. https://doi.org/10.1007/s11581-023-05205-6. <Go to ISI>://WOS:001070490900001.

Zhu, Xufan, Wei Wang, Guoping Zou, Chen Zhou, and Hongliang Zou. 2022. "State of Health Estimation of Lithium-Ion Battery by Removing Model Redundancy through Aging Mechanism." *Journal of Energy Storage* 52. https://doi.org/10.1016/j.est.2022.105018. <Go to ISI>://WOS:000814722500006.

Zhuang, Quanchao, Zi Yang, Lei Zhang, and Yanhua Cui. 2020. "Research Progress on Diagnosis of Electrochemical Impedance Spectroscopy in Lithium Ion Batteries." *Progress in Chemistry* 32 (6): 761–791. https://doi.org/10.7536/pc191116. <Go to ISI>://WOS:000577098200007.

Ziegler, Andreas, David Oeser, Thiemo Hein, Daniel Montesinos-Miracle, and Ansgar Ackva. 2021. "Reducing Cell to Cell Variation of Lithium-Ion Battery Packs During Operation." *IEEE Access* 9: 24994–25001. https://doi.org/10.1109/access.2021.3057125. <Go to ISI>://WOS:000617754700001.

Zihrul, Christiane, Mark Lippke, and Arno Kwade. 2023. "Model Development for Binder Migration within Lithium-Ion Battery Electrodes during the Drying Process." *Batteries-Basel* 9 (9). https://doi.org/10.3390/batteries9090455. <Go to ISI>://WOS:001076820100001.

Zitouni, Farouq, Saad Harous, and Ramdane Maamri. 2021. "A Novel Quantum Firefly Algorithm for Global Optimization." *Arabian Journal for Science and Engineering* 46 (9): 8741–8759. https://doi.org/10.1007/s13369-021-05608-5. <Go to ISI>://WOS:000645472700001.

Zou, Dongyao, Ming Li, Dandan Wang, Nana Li, Rijian Su, Pu Zhang, Yong Gan, and Jingjing Cheng. 2020. "Temperature Estimation of Lithium-Ion Battery Based on an Improved Magnetic Nanoparticle Thermometer." *IEEE Access* 8: 135491–135498. https://doi.org/10.1109/access.2020.3007932. <Go to ISI>://WOS:000554910400001.

Printed in the United States
by Baker & Taylor Publisher Services